Bee Better Beekeeping

A guideline for the sustainable honey bee keeping and care of healthy honey bees in a responsible, thoughtful, studied and easy way

Herbert ten Thij

IBRA
INTERNATIONAL BEE RESEARCH ASSOCIATION

Bee Better Beekeeping

Jointly published by: The International Bee Research Association,
a Company Limited by Guarantee, 1, Agincourt Street, Monmouth, NP25 3DZ (UK) &
Northern Bee Books, Scout Bottom Farm, Mytholmroyd, Hebden Bridge HX7 SJS (UK).

Obtainable from:
www.ibra.org.uk & www.northernbeebooks.co.uk

Published 2026 by IBRA - International Bee Research Association and
Northern Bee Books,
Scout Bottom Farm,
Mytholmroyd,
West Yorkshire
HX7 5JS (UK)
Tel: 01422 882751
Fax: 01422 886157
www.northernbeebooks.co.uk

ISBN 978-1-913811-19-8

Illustrations: if not mentioned otherwise: Drs. Herbert A.C. ten Thij

Design and artwork DM Design and Print

Index

Preface

This book is about beekeeping as it has been developed in Germany by Dr. Gerhard Liebig (Bieneninstitut Universtät Hohenheim) and then also by Dr. Pia Aumeier and thirdly also adopted by Dr. Otto Boecking, Bieneninstitut Celle, et al.

It is beekeeping based on knowledge of facts that have been acquired in decades of research or thousands of hours of study, investigations and testing. This has not only happened in hundreds of hives, in apiaries in different climate zones, but sometimes also with the help of beekeepers nationwide to obtain as much data as possible from as many different circumstances as possible. In this way many a traditional truth and hypothesis has perished. All to the level of knowledge of the beekeepers and the health of their honey bees.

At the same time the beekeeping method has been developed the way it is least time consuming. If one has to take care of several hundreds of hives in several apiaries this is a necessity. All this in the end to the advantage of the beekeepers (less effort, less work and easy keeping of honeybees).

Thanks are also due to Dr. Gerhard Liebig for making also his photographic material available, to Mr. Stuart Roberts, International Bee Research Association - IBRA, for his language advice and suggestions to make the Euro-English as provided by me readable and, in that perspective, to Prof. Jamie Ellis, University of Florida, and Dr. Bram Cornelissen, Wageningen University, for advice on the translation of Dutch terms that strangely have no equivalent in English.

Markelo, 2025

Herbert ten Thij started beekeeping with succesfully completing a course about eleven years ago in Enschede, the Netherlands at an affiliated association of the Nederlandse Bijenhoudersvereniging - NBV [Dutch Association of Beekeepers] while he had a partial domicile as well in Germany. He started his apiary in Germany and so he also joined the Imkerverein Gescher-Stadtlohn-Vielen u. Umgebung e.V. [the Association of Beekeepers in Gescher-Stadtloh-Vielen and surroundings, Germany*]. Having a focus also on beekeeping methods in Germany he became interested very much in the scientific based way of beekeeping as developed by Dr. Gerhard Liebig and Dr. Pia Aumeier. He joined the Dr. Pia Aumeier's courses/practices 'Making Queens in a Team', 'The Art of Making Wax Products' and 'Becoming an Expert in Moving Hornets and other Wasps'. Over time he became a referent of Mrs. Dr. Pia Aumeier organising/giving the affiliated course 'Making Queens in a Team' and also since 2023 in his new home in Markelo, the Netherlands. He gives annual courses in beekeeping according to the Liebig/Aumeier method as well.

He is at present a member of the Beekeepers Association "Eemkwartier" in Amersfoort, the Netherlands of which he has been a member of the Board till the times of corona. The same counts for ending his membership of the Study Group of the NBV. During a summer holiday in 2018 he visited beekeepers in Liguria, Italy to study their ways of beekeeping in a climate that sooner or later will be the same in our regions. He is a member of IBRA – the International Bee Research Association as well.

Before writing this book he translated the fourth edition of the book of Dr. Gerhard Liebig from German into Dutch and published it. The book was sold out within half a year. While next being busy in translating course material for an expected English speaking Mexican student in 2025 he contacted Mr. Stuart Roberts of the IBRA for consultations of translating some beekeeping terms unknown to dictionaries Dutch-English. This way he was invited to write this book.

* Should you ever be in this region do also visit the "Westfälisch-Niederländischen Imkermuseum"/"Bienenmuseum Gescher" [Westfälian-Dutch Beekeepers Museum / Bee Museum] run by this association. It has also a nice collection of amber inclusions of 'prehistoric' bees.

Bee Better Beekeeping

Herbert ten Thij

The famous French writer Gustave Flaubert tells in his notebook *"Voyage au Oriënt"* about dances he admired in Egypt at that time. The nature and intention of those bee dances can be guessed at, but what the dancers exclaimed is crystal clear: 'In ny a oh! In ny a oh!' 'Beware of the bee! Watch out for the bee!

There are many bees that have lost the ability to sting over the course of their evolution, but not our honey bees. Well, as for the workers and the queen, because the males - the drones - they don't sting. Fortunately, neither the fact that honey bees can sting nor the old knowledge of it does have to stop us from enjoying beekeeping. Even already a little smoke keeps the bees and also the beekeeper calm. Almost everyone who starts beekeeping is initially shy or even afraid for these little animals and will work with them and in his or her hive and colony with caution and respect. But with experience comes confidence.

Once the beginners have learned how to deal with bees, they will behave differently than when they first contacted them.

In the beginning, every beekeeper will protect themselves well when he or she visits the bees. They will put on a bee hood or veil and may also put on leather gloves. There's nothing wrong with that in the beginning. However, it should not remain that way.

Only if you have an allergy to bee stings is it advisable to avoid being stung. Then and only then is continuing to wear protective clothing a permanent necessity.

There is usually little risk of stinging if the beehive is approached calmly and the smoker is at hand. However, when visiting bees for the first time, it is good to keep an eye on what happens when a sting has occurred (and most certainly when there is a risk of an allergy).

Bee sting protocol (for allergies) - what a beekeeper or a visitor to an apiary with a possible allergy can do in the event of a bee sting (mostly for the first time)

1. Remove the sting as quickly as possible.

   (Usually there is still a venom bladder attached to the sting that does not stop pumping and so injecting. A bee that has stung a person dies because the sting anchored in the skin tears the bee apart, but the venom bladder stays attached to the sting most of the time)
2. If possible, communicate that you have been stung.
3. Stay then within sight of other people. The other people should continue to observe the person who has been stung.
4. Take complaints serious.

   Complaints:

   a) swelling at the site of the sting. This may also only occur later.
   - local itching
   - local redness
   - local swelling
     - possibly use an anti-histamine tablet

   b) skin symptoms all over the body.
   - hives, nettle rash, urticaria,
   - itching
     - antihistamine, go to the doctor, prednisone

   c) general redness, swelling and gastrointestinal symptoms.
   - swelling in other places
   - intestinal complaints
   - abdominal pain
   - nausea
   - vomiting
   - diarrhea
     - antihistamine, go to the doctor, prednisone

   d) respiratory symptoms.
   - shortness of breath, wheezing, hoarseness, tight throat, shortness of breath

     drowsiness, fainting
     - call the emergency number, use an EpiPen if there is a danger to life

e) vascular disorders, anaphylactic shock
- palpitations,
- chest pain
- drop in blood pressure
- urine loss
- blue hands/feet
- unconsciousness
  - call the emergency number, use an EpiPen if there is a risk to life

5. If you have complaints b) and c), go to the doctor. The affected person is not allowed to drive. If necessary, use antihistamine tablets.
6. Call emergencies care (emergency number) for complaints d) and e). Notify emergency room or medical provider that an EpiPen will likely need to be used.
7. Take precaution: have a telephone and the emergency number at hand, as well as the address of the apiary and anti-histamine tablets.

You will probably never have to use this sting protocol in your future life with bees, but for the sake of completeness and certainty, it is good to know that it exists.

As far as your own attitude towards the bees' stings is concerned, that is (also) more a problem between the ears than between you and the bees... After a while the beekeeper will develop a feeling for his bees and he or she knows how to behave to avoid problems.

Make sure that the smoker is on when you visit the bees and that you have a veil on hand in case you encounter a stinging problem.

You grab the veil only to put it on when necessary.

The queen of any stingy colony is removed and replaced by a gentle queen, whether or not produced by the colony itself.

Only put on gloves in exceptional cases, if the bees of a colony are very aggressive and working in the colonies cannot be suspended and is absolutely necessary.

The use of gloves

The advise in the UK is to use a pair of thin disposable latex or nitrile gloves rather than no gloves. They offer no protection from stings but keep the spread of disease at bay. Probably you have to change gloves for every visited hive for all certainty. Anyway they keep your hands clean from honey or propolis. In Europe mainland the believe is that with gloves it is difficult or impossible to develop the sensation required for handling bees. This applies in particular to raising and replacing the bee frames. Of course there are exceptions, for example for those beekeepers who have or have developed a serious allergy to bee venom. This rarely occurs.

If a bee sting regularly causes physical discomfort, circulatory disorders, severe swelling or skin rashes, the beekeeper has no choice but to work well protected in his colonies. Even in normally very gentle colonies one can occasionally get a sting, for example if one hurts a bee removing or hanging frames. Beginners in particular react very sensitively to the first stings. As I remember well when I got my first sting in my 'construction workers cleavage' when I bent over and watched as a fellow student clumsily lifted a frame. However, a sting in the eyelid is enough to cause the eye to swell so much that one or two days one can hardly see anything with this eye. A stab in the hand can cause the entire arm to swell up to the shoulder.

Such experiences are painful, but no reason to give up beekeeping. Generally it usually leads to such habituation that stings can still be felt for a few seconds and may still hurt a little, but otherwise no longer evoke any physical reactions. Your body adapts.

Nevertheless, as said before, especially during the first stings, a victim must carefully monitor his or her physical reactions and his/her well-being. Again, a circulatory system that runs into problems or even stagnates is not to be trifled with. If a bee sting is life-threatening, beekeeping should be discouraged. Anyone who cannot or does not want to do it should consult his doctor. Desensitisation can and has been achieved with medical treatment and already helped many beekeepers. Such cases are very rare. As a rule, the body reacts violently only to the first sting or stings. It builds up antibodies against the bee venom and becomes immune. After that, bee stings do not hurt so much anymore for most colony or only hurt a little.

What is the trouble with veils or bee hoods?

Less than against gloves. But bee hoods or veils also hinder the work. They easily slip off the head in windy weather and make it harder to see through them. One bee hood is not the same as the other. The first bee hood I had was part of a beautiful white coat. Unfortunately my head was too small for this lid with veil. In that way I was more concerned with the bee veil than with the bees. For a better view, it may be worthwhile to make a simple veil yourself. Heavy, fine tulle curtain fabric is sufficient as material. Stings can also be prevented in ways other than wearing protective clothing.

Pay attention to the mildness of your colonies!

In principle, one should keep only "gentle" colonies, not only out of self-interest, but also in the interest of family and neighbours. If a colony stands out for its above-average aggressive behaviour, which is manifested, for example, if the beekeeper is attacked by the bees when opening the colony or, even worse, when approaching the colony, then the queen is replaced on occasion by another queen. Of course, this measure rarely helps immediately, because it takes two months before the colony consists entirely of descendants of the new queen.

When choosing propagating material for raising queens for colony propagation and for queen replacement of production colonies, the gentleness characteristic has priority. Only gently colonies are bred.

Prepare well for working in the bee colony!

What do I want to do? All things such as the protocol cards (or pen and paper), tools, frame material and other items that will be used or are necessary for the expected work in the colony must be ready at the colony cabinet stand before the start of work in the colony, so that the colonies need to be opened for as short a time as possible and only once.

Bees have an excellent sense of smell and can react to body odours, which should be taken into account when using perfume, soap and shampoo.

The smoker must burn well and not go out after the first puffs of smoke. Suitable as smoking material is, for example, dried wood, shavings, wooden firelighters (for an open fire - to light a fireplace) or straw pellets (bedding material for rabbit hutches or guinea pig stables) which usually I use. For lighting, a piece of egg box or crumpled kitchen paper is used. I use myself a (soldering) gas burner to light immediately the straw pellets, quickly and easily, but one should be careful that only the material in the smoker starts to burn and smoke...

Only put the smoking material in the smoker when the egg lace or paper towel is 'on fire'. The smoker is after working in the colonies suffocated with a tight wad of kitchen paper, because there is a risk of a 'forest fire' when emptying the still burning or smouldering smoking material! I place the smoker on its side and/or press a tuft of fresh, wet grass into the smoke nozzle. After a few minutes the fire will be extinguished and you can store the still hot smoker when it has cooled down sufficiently.

That is why a bottle or barrel of water should also be part of the standard equipment of a beekeeper. With this, the glowing contents of the smoker can certainly be extinguished, the honey-sticky hand brush can be cleaned and one can wash one's hands before getting behind the wheel of a car again after work in the countryside. A good way to remove propolis from hands is to rub them with methylated spirits. Be careful of course because methylated spirits is a fire hazard.

Work quietly

Hectic activity in the apiary must be avoided. The beekeeper's behaviour becomes easily taken over by the bees. Above all, the beekeeper must maintain inner peace. Nervous hand movements should be avoided. Bees prefer to fly on anything that shines (jewellery, eyes, teeth - anything that resembles the eyes of a bear or other aggressor that hunts for honey and larvae). So, it is therefore advisable not to wear jewellery or watches on your wrist and to keep your mouth closed when working in the colony. Bees often attack suddenly. The flying and stinging sometimes happens in an instant, unless an attacking bee gets lost in the hair on the head. Then you still have some time to grab the bee and remove it before it stings. But of course you have to learn it first (you can practice with drones) how to grab a bee with your fingers without getting stung...

When the bee lands on bare skin, the defence is usually too late. Then nothing remains otherwise more left than removing the sting. In any case, if you are attacked and have a frame in your hands, you should not drop it. It would also be wrong to hit the bee or bees when they keep coming flying. That would only enrage them even more. It is better to behave calmly, not to hit at it, to put on a bee hood or to slowly move away and find a somewhat secluded 'hiding place' searching behind a tree or bush or behind or in the car (if you drove to the apiary)...

If the colony you have just worked in remain open for a few minutes, the aggressive bees notice, after they have chased the beekeeper to flight, upon their return in the colony that they have a much bigger problem to solve there: the regulation of it's climate. On his return, the beekeeper purposefully uses more smoke to complete the work in the colony.

Avoid robbery!

Every robbery is preceded by a mistake by the beekeeper. During periods of low flow, especially in late summer, urgently necessary work in a bee colony is best completed before the flight of the bees can be initiated. Robbery quickly turns into stinging and sometimes only fades away afterward to dawn. Once robbery has been unleashed, it occurs more quickly on the next visit to the apiary. The colonies themselves have to fear this, and in particular the weaker colonies, whose entrances must therefore in principle be reduced. If robbery has broken out, reducing all entrances and fly gaps prevents the loss of colonies. On the other hand, these measures can be ignored if one has not committed any errors. The mistakes you can make are discussed below.

Don't open a hive without smoke

With meek colonies you can certainly look in from above or even reach frames without being stung. Nevertheless, in principle you should have a smoking smoker ready to grab immediately and deploy in time. As said before: smoke calms the bees and the beekeeper too!

Watch the mood of the bees

Bee colonies are not constant in their behaviour, but just like the beekeeper, they are subject to moods. If there is no nectar flow, all the bees are at home, the flying bees hang around idly. They are easily irritated. In addition, there are scouts from every colony on the road who are constantly investigating whether anything can be obtained from their neighbours. This generally increases the willingness to defend, which the beekeeper may also have to deal with. When there is flow flight, when the bees are busy collecting nectar, the colonies are easier to work and there is less "danger of stinging". Perhaps this is due to the fact that the older bees are on the road and not in the colony, because they are more likely to come up with ideas than the young bees in the colony. During (nectar) flow there is also plenty of work to be done in the "indoor shift", because all the bees in the colony are busy processing the nectar or honeydew carried in and have their bellies full. That too inhibits the desire to sting. The effect of the smoke is also explainable. The bees pass through the smoke to gorge themselves on honey. That takes a few minutes. Then the honey bladder is full and the bees are satiated. Satiated bees do not sting. Of course, the bees can only be calmed with smoke if open food is present! If that is missing, fumigation with the smoker will not work in the long run. The bees are only driven away for a moment. A few flee from the smoke and then attack the disrupter.

Save working in the most restless colonies for last

The observant beekeeper knows his colonies. If a colony repeatedly stands out for aggressive behaviour and the causes cannot be remedied immediately, such a colony should always be processed last. Because the closure of the colony does not stop the attacks of aggressive bees. They often follow the beekeeper until he or she leaves the apiary.

Searching for a location

A lady once asked me to put a beehive in her garden. According to her, it would contribute to more biodiversity. It is not for me to pass judgement on that. I just placed a beehive in her garden. After a few weeks she surprisingly started saying that the bees are always gentle with me, but are always aggressive towards her... whenever she was working in the garden. She was constantly pruning and moving plants around the hive. She had no idea that all that digging and hurting plants (although everyone knows, for example, how freshly cut grass smells: "alarm, alarm") give the bees the impression that the big bad beer is coming or some other honey and larvae snacker. This example makes it clear that it is important to carefully consider where the beehive will be located. There is however more to take into account when choosing a good location, to find a good spot for the hives. Indeed: hives, because it usually does not stop at just one hive. Most beekeepers have four or more hives after a few years.

There must also be a place for empty containers, for the tools, for frames, foundations and combs, and for the honey extractor, the honey sieves, the hair dryer/uncapping tools, the honey buckets, the supply of fondant, syrup or sugar water and a wax smelter. This is usually part of a utility room or a shed.

Also you have to take in account what your neighbours' attitude will be when they will be informed that a beehive will be standing next door. There is the risk of a sudden increase in new allergies on the street if the news gets out and the beekeeper isn't that popular in the neighbourhood.

If the bees sting the beekeeper, it is less serious than if it happens to someone else. Remember well that every animal owner is responsible for his animals. Sometimes bee stings can have unpleasant consequences, as seen in the bee sting protocol. So, moreover if the person who was stung needs to seek medical attention. It can happen, although very rarely, that bee stings lead to death. In any case, the beekeeper must be responsible for the consequences. Anyone who is a member of a beekeepers' association and pays contributions is liable for very favourable conditions. For that reason alone, it is worthwhile to be a member of a beekeeping association. So is in the

UK by membership of the BBKA (British Beekeepers Association) the member covered for £15M for such cases. But even then it is certainly better not to have to fall back on liability insurance. The best precaution is to arrange the colony in such a way that no one is in danger. This must be taken into account when looking for a place bees fly in and out of the entrance. There is a lot of air traffic, especially when the flow is good. This air traffic can be controlled by the arrangement of the hives or by appropriate planting in the vicinity of the beehives. A fence a few meters in front of the entrances or between the bee stand and the neighbour's yard or garden directs the flying bees upwards or prevents the bees from flying at eye level with the neighbours. A hedge or fence directly behind the apiary also ensures that the beekeeper or his guests are not in the flight stream even if the flow source is behind the apiary. For years there was a beehive behind a hedge in a small city garden. The children playing in front of the hedge were never stung. A striking example of how effective the fence or hedge measure is.

The bees returning home from the flow or the bees that fly out in search of food sources are not aggressive and rarely sting. Nevertheless, no one should be directly in front of the entrance, especially when the beekeeper is working in the colonies. Behind the colonies the danger of stinging is clearly less than for the entrance side.

Beekeeping should remain a pleasure

Keeping bees means, among other things, dragging hives and frames. When the combs are filled with honey it can be very tiring to carry them. Therefore, colonies should be arranged in such a way that tendons and muscles are not unnecessarily stressed. Keep in mind that also you are getting older. Steep slopes should be avoided when setting up the bee colonies. The most suitable places are those that you can reach by car or transport cart up to in the apiary, regardless of the weather conditions at the time or before.

The location should be good for bees as well of course

Bees can fly, fly far away. Therefore, the requirements for a bee stand do not have to be met in the immediate vicinity of the stand.

The suitability of a site depends on the flow situation, in particular the nectar and pollen supply in spring and late summer. It can be considered good if there is a lot of varied greenery in the nearby flight radius (approximately 2-3 km).

Particularly favourable are extensive wild meadows, fruit orchards, small gardens and front gardens, hedges, tree- or hedge-lined streams, parks, avenues and mixed woodland.

Interested city residents may have reservations about keeping bees because they lack space and a garden or yard. But that should not prevent you from becoming a beekeeper if you follow the recommendation: start small and modestly! City beekeeping has even become a custom and a trend nowadays.

Colonies do not need their own garden. Your own garden cannot be the most suitable place either. Consult with a forester or with farmers who own meadows on forest edges. Hedges, slopes or even better, forests that are in front of the colony block strong winds, guide the bee flight upwards and are at the same time a food source for the bees. It is also advantageous if the bee colonies are placed under deciduous trees. So, the hives are not in full midday sun in the summer and in the winter and spring they can also benefit from the sun's heat that falls on the flight slit or the flight board in the afternoon.

If you only keep three to six colonies, a cellar or shed could suffice initially to store beekeeping supplies. It is of course easier if the colonies are next to the garden house where all beekeeping supplies are stored and are quickly available. If you don't have all that but want it, you have to invest a lot. In addition to the investments, one must also pay attention to the ongoing costs, which increase with the growing colony populations and burden the money bag every year if beekeeping success (= honey yield, selling queens or colonies, selling candles or other wax products) does not occur.

The line-up of the colonies

The conditions of the location determine in which direction the entrance will be located. On a slope, e any difference for the bees where they are placed, you can make than your own interests take precedence. Most beekeepers prefer shady and wind-protected places.

Best house the beehives in the open space

A bee house is expensive. Moreover, the work in the colonies is in a bee house more laborious. It is better to house the colonies outdoors and this more over when they are kept in hives with boxes. The colonies are placed on a scaffold (a trestle). For this purpose, a pallet that can accommodate two colonies is sufficient. A trestle can also be made with two beams. To prevent the trestle from rotting so quickly, the pallets or beams are placed on stones.

A landing site at the entrance is actually unnecessary. The plant growth in front of the entrance is kept short manually.

If the colonies are arranged individually or in pairs, they can also be worked on from the side. Then caring for the colonies during the course of the year is easier than with an operation that can only take place from the back or if many colonies are close together.

The materials needed / the investments to be made

Keeping bees costs money. A range of products is offered in the beekeeping trade. In the beginning one must limit oneself to the essentials. Even later, when you have gained enough experience in dealing with bees and your number of colonies increases, you will not need all the equipment that the market has to offer. The following overview shows you what you would need in the beginning:

per colony

1 simple bee hive of three boxes *of the same size*
(also when choosing a British Standard National Hive choose than also one built up with 3 boxes -all of equal size- yes do so, if you prefer easy beekeeping. Probably the best is than to choose the Standard version) with 30 wired frames (preferably with Hoffmann sides. Since in the UK many beekeepers use already framed foundations than frames tat have been adapted accordingly of course), mesh bottom box,

inner lid and
metal roof)
1 queen excluder
3 kg foundations (wax plates)
of course 1 bee colony and
20 kg of sugar

for the work in/with the bee colony

1 smoker (for your comfort, use a smoker
with a diameter of at least four inches - ten cm)
A Dadant type (see figure) is handy.

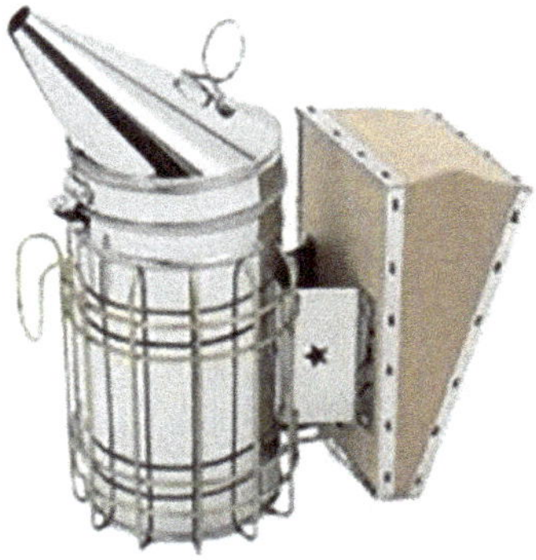

1 hive tool
and
1 bee brush
a beekeeper hood (veil)
a few tension wires

for the honey harvest

1 three- or four-frame honey extractor made of stainless steel, with or without motor,
1 uncapping equipment (uncapping fork) (hair dryer)
1 honey sieve set (2 stainless sieves)
12 kg clean or better new plastic honey buckets (buckets that are correct for storing foodstuffs)

for comb hygiene

1 wax melter (e.g. a kitchen sink that fits into a beehive box)
1 vapour device (e.g. a vapour machine for removing wallpaper)

for the honey harvest

1 three- or four-frame honey extractor made of stainless steel, with or without motor,
1 uncapping equipment (uncapping fork) (hair dryer)
1 honey sieve set (2 stainless sieves)
12 kg clean or better new plastic honey buckets (buckets that are correct for storing foodstuffs)

for comb hygiene

1 wax melter (e.g. a kitchen sink that fits into a beehive box)
1 vapour device (e.g. a vapour machine for removing wallpaper)

It is advisable to first check the prices in the trade - including internationally... - before purchasing materials or supplies. When ordering via the Internet, also pay close attention to the shipping costs.

Which hive?

New hives are more expensive than old, used ones. It is not surprising that a beginner would consider purchasing hive materials from another beekeeper who may be giving up beekeeping due to old age or reducing his number of colonies. But that is often a big mistake. A mistake that is made by many beginners. Unfortunately, I made that mistake too, buying a hive from a beekeeper who emigrated to Spain, and I only ended up at the "right hive" after two years of misery!

There are a multitude of hive forms and frame types in use. A look at the advertising sections of (at least the continental) beekeeping magazines makes this clear month by month. There are even more types available on the Internet. Accepting an allegedly favourable offer can easily lead to a mistake. Such a mistake is completely guaranteed if one decides to use bee houses that are recommended for suspect or marketing reasons. Recommended beehives as being especially suitable for the bee species at hand, as true to nature or as suitable for organic beekeeping or bio-beekeeping, such as the "bee box", the "bee bobbin", the "top bar hive" hive, the Warré hive or round bee enclosures that simulate a tree cavity.

In principle, bees can be kept in any hive. That is in itself not the problem. Some experimentalists have even kept bees in big flowerpots. This also makes clear the multitude of hive types in circulation. But not every hive is economical, that is to say, cheap and efficient at the same time. For example a Dadant hive or a one box hive takes 59% more work (as calculated by Dr. Pia Aumeier from numerous tests) than honybees kept on two brood boxes (than the tilt control is possible). Also not every hive is suitable for easy and responsible beekeeping.

It is of course also possible (and very easy and much cheaper) to make the boxes of the bee hive yourself. A construction manual you may find as at the end of this book.

Preference of a hive with boxes

On the continent and foremost in the German-speaking countries, for many years the hive that had to be worked from behind (and of which there are almost countless constructions) dominated for decades. With most types, the space available to the bee colonies is limited to one breeding area and one honey area.

In these hives the colony hibernates in the breeding area. In the spring, a honey chamber on top is made accessible. The colony in such a hive is cared for exclusively from behind. Although the brood frames can be lifted (slid out) without having to remove the brooding space, it cannot be taken off either, because the brooding space and honey space are in one closed hive. The bee colonies can be stacked, the beekeeper's access from the rear makes this possible. Something like this is often necessary of course if the bee colonies are housed in a respective bee house.

In a hive with separate boxes, the colonies are cared for from above. That is considerably easier than taking care of the bees from the back. The honey chamber is placed on the box or boxes of the breeding area above a queen extruder. The honey chamber can consist of one or even three boxes, depending on the strength (size) of the colony and the availability of nectar flows. The colony can be enlarged from above at if desired so.

Colonies that are box-composed can be placed in the open field. A bee house is not necessary. Every colony lives in its own house.

The increased spread of hive boxes in the second half of the twentieth century has not led to a uniformity of hive types on the continent.

On the contrary. Just like with the hives that need to be worked from the back there is now also an incredible amount of constructions and self-build variants for box hives. Many of them are so complicated that it is hardly possible to build your own hive at an affordable price. They often consist of many parts and are therefore very expensive to purchase.

In the Netherlands, a type of hive is still popular that is approximately 8/10 the size of a Zander sized or about a Langstroth sized hive and in which the honey space is usually a half box with separate, half-size windows. Not all that convenient.

It is recommended to buy or build a beehive with one type of boxes. This way the boxes are interchangeable. And that is certainly useful, but certainly also in the control strategies against parasites.

- boxes of wood (preferably light white pine wood)
- the same frame size for brood box and honey box (Zander or Langstroth format)
- space for ten frames per box
- for cold built hanging of the frames

- with walls 2 cm thick

- boxes without a notch
- handle bars instead of recessed handles
- no seizures
- handle bars instead of recessed handles

box front or back

- floor = mesh bottom box (of sufficient height: min. 8 cm)
- at the back: opening to push in and out the inspection board

front mesh bottom box (floor)

- with queen excluder
- The example in the picture is a plastic version.

  There are also metallic versions.

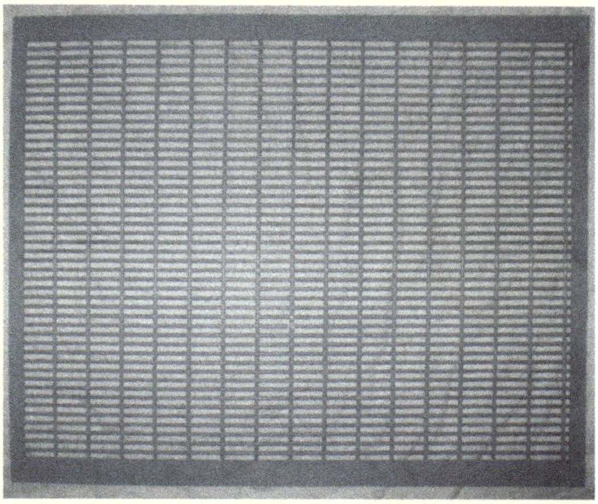

Boxes made of wood

It does not matter at all for the development of the colony whether they are housed in wooden or plastic boxes. Plastic boxes insulate better, which is only reflected in slightly lower feed consumption. For comb hygiene, it is absolutely recommended to keep the colonies above an open mesh bottom box all year round, especially in winter and when spring arrives in time. Then it doesn't really matter if the walls are well insulated.

The parts of a hive consisting out of boxes are shown at the front:

mesh bottom box, with the entrance (the floor)

two brood boxes, (lower - and upper brood box)

queen excluder (on top of the upper brood box),

one (or two or even more) honey boxes,

the foil and inner lid are not visible as they are hidden under the metal roof

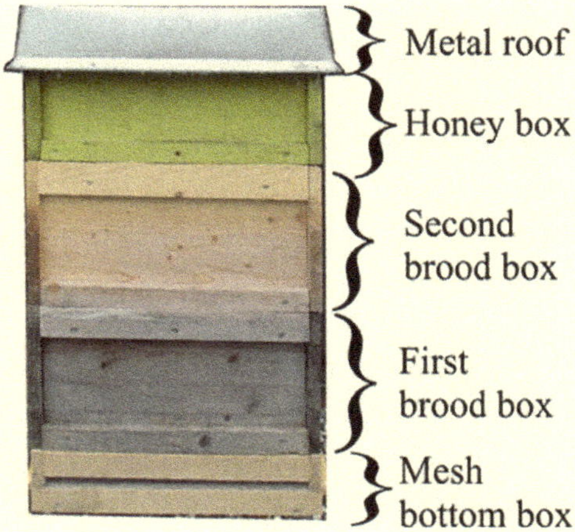

Plastic containers are lighter than wooden boxes. Of course, they also cover more volume. They need more space on a trailer or lorry. This point of view only comes into play when you have multiple colonies and you travel with them. Nevertheless, one must pay attention to it from the start. The decision for plastic or wood is primarily a matter of attitude. Beekeepers pride themselves on the fact that beekeeping makes a significant contribution to nature conservation and the protection of the environment. As a nature conservationist, one should not use materials that pollute the environment during production and waste processing. In addition, the natural housing of the honey bee colony, the hollow tree trunk, consists of wood.

You can build wooden boxes yourself if you have an adequately equipped workshop. Or in the kitchen if there is enough tolerance in your household. If the woodpecker has struck during the winter, the damage to wooden hives is usually less than to plastic ones and also easier to repair. With woodpecker repellent installed in time, before the first snow falls, wooden and plastic hives hibernate undamaged.

In general, one cannot give preference to a particular type of wood of the beehive boxes. It depends on what requirements the beekeeper places on his hive. If a large company has adequate lifting vehicles or equipment that make the transport of hives and colonies easier, then using pine wood, for example, certainly makes sense, because it is very stable and durable, but also relatively heavy.

Beekeepers as a secondary profession or hobby beekeepers, however, better use the lighter wood of the white pine (Weymouth pine wood). To make this light wood stand out, the cabinet walls should not be too thick. Two cm thick is more than sufficient. The quality of the wood naturally has a decisive influence on the shelf life and stability of the beehive. The wood should not have blue or red stripes, because mold growth attacks the wood and shortens its lifespan. In principle, one should work with knot-free planks. With white pine wood, having knots plays a lesser role. The moisture content of the wood should be approximately 15%.

The size of the boxes

Which boxes are used depends greatly on their availability. For example, at the time I was interested in a beehive in the Zander size, it was not available in my country. I therefore bought my first Zander box beehive via the internet in Bulgaria. To have an example and a comparison of the sizes for the boxes I would build myself.

Which box is chosen can also depend on the choice of the frames that will be used. See hereafter about the choice of frames.

The Zander box hive was chosen because of the ease of estimating bee numbers and numbers of worker brood cells to monitor the development of the bee colonies during the year. Also the weight of the honey combs (ca. 20 kg) or the honey chamber has been considered. In addition, Dr. Gerhard Liebig and Mrs. Dr. Pia Aumeier having tested many different bee hives advice to use this type of bee hive. Also in the long run this turned out to be a useful advice.

But again: for the bees it makes no difference at all which hive is used. The difference between hive types is made by what suits the beekeeper best.

A few for comparison:

Frame size, breeding and honey chamber (Hoffman type)	'1' in mm	'2' in mm	'3' in mm	'4' in mm	'5' in mm	'3' ✣ '4' in cm✣
Zander	420	220	400	200	28	800
Langstroth	448	232	428	203		869
BSNH* (BSHF)	356	215			38	
BCDF** (16"x 10")	404	264	386		17	460
Dadant	435	300				

Boxes (frame capacity - FC: 10)					
Boxes (10)	*Length:*	(interior)	*Width:*	(interior)	*Depth:*
Zander	520 mm	450 mm	414 mm	380 mm	230 mm
* British Standard National Hive (FC: 10/11)	460 mm		460 mm		225 mm
Langstroth	505 mm	465 mm	405 mm	365 mm	238 mm
Dadant Blatt	500 mm		430 mm		310 mm

Frame capacity	bees	worker cells	drone cells	food
Zander, per frame side	1,000	3,200	1,840	1,000 gr.
Simplex, German Normal, etc.: ca. 80% of these numbers				

* BSNH: British Standard Hoffmann Frame

**BCDF: British Commercial Deep Frame

See for the choice of frames hereafter.

The tilt control

From the beginning of May on, the bee colonies must be checked every seven days to see whether they are in swarming mood. The seven-day rhythm is also reflected in the mnemonic "three, five, eight and the queen is made". It reminds us of the development period of sixteen days of a queen. Three days egg, five days larva and eight days pupa. The seven-day rhythm must be strictly adhered to during the swarming period so that the swarming mood is kept in check and controlled at the apiary. Swarming must be prevented.

The swarming period, for that matter, ends after the longest day.

During the tilt check, the two breeding boxes are separated by tilting the top box upwards and at the same time pulling it slightly forward. This makes it easy to see whether swarm cells have been created in the bee colony. If that is the case, they must of course be removed. If necessary, a flashlight can be used for a better look.

Swarm cells are easy to recognise and they are usually built at the bottom of the frames. If there are swarm cells, each frame must be checked. If no swarm cells are visible during the tilt check, then, as years of experience with hundreds of colonies have shown, there is hardly any chance that there are swarm cells in the bee colony.

Thus, the tilt control helps make beekeeping easier and less time-consuming. It is a 95% reliable swarm control method.

A tilt control carried out in winter. The bees are huddled together in a ball that is now divided in half by the tilt control. The bee cluster is close to the entrance. (Picture after Dr. G. Liebig)

A notch make tilting boxes difficult

Boxes [containers] with a notch complicate the connection between the boxes, the mesh bottom box and the lid parts. The notch ensures that the mounted boxes do not shift, but this can also be achieved by placing a sliding bar under the short ends of the boxes. The notch makes the so-called tilt control difficult, which is frequently required during the swarming period in May and June at a distance of seven days.

During the tilt check, the breeding space of a colony is split, the upper box is pushed forward a little and then tilted upwards. In this position, the brood frames of the upper, tilted tank can be viewed very well from below to see whether swarm cells have been produced there. If the test is positive, each frame of the breeding space must then be lifted and checked for swarm cells, which are then destroyed. If the test yields negative results (the colonies must be cared for in the spring in such a way that this is normal), the colony will be closed again. If the boxes have a notch, the top bin must be lifted over the notch in order to tilt the bin. This makes tilting more cumbersome and requires strength. That is why you should not purchase or make boxes with a notch.

Grip bars

The boxes are multifunctional. They are also used for storing frames (in which case an extra frame box is not necessary) and for their transport. When the combs are full of honey, the boxes are very heavy. A box with 10 full honey frames (Zander) weighs about (20 + weight box, combs and frames) kg. Such boxes are easier to handle and carry if they are equipped with wide handle bars. In recessed grips, hands slip and fingernails tear or break easily.

Better no batter

Many beekeepers experience the removal and re-installation of heavy honey boxes as heavy labour and this is a reason why older beekeepers in particular are adverse to beekeeping with hives with boxes. An attempt has been made to tackle this problem with the development of tilting hives. Their boxes are equipped with hooks, locks and hinges, which makes the construction of those boxes more complicated. A well-secured hook is also required to tilt the honey bins forward. The hives must be firmly anchored to the trestles, otherwise there is a risk that the colony will fall over when the honey tank is tilted away. Usually trestles are used on which five or more colonies can be placed. This in turn has the disadvantage that all colonies that are on the trestle are "awakened" as soon as the work in the first colony is started.

Use a mesh bottom box as a floor

Actually, just for all clarity, this part should have been titled: Use a Mesh Bottom Box Instead of a Bottom Floor. The mesh bottom box (container) is much more suitable for keeping bees than a solid bottom. Because:

- it makes beekeeping with the hive easier. No combs will mold if the colony sits above an open mesh base all year round, including winter and spring. When overwintering in two boxes, the unoccupied edge frames can remain in the colony; they do not have to be removed in the autumn after feeding or replaced in the spring when the bees start building.
- the colonies, placed close to the ground or on a low trestle, can be kept on an open mesh bottom box all year round without their development being disturbed at any time, not even in the spring when there has been a persistent cold wave for a long time. That does not harm the colonies if they have adequate food supplies. They always depend on this, for that matter, regardless of which hive or on which floor they are kept.
- in the era of varroa mites, each hive must be equipped with a mesh bottom box that can be closed from the back (and from below) with a sliding drawer (counting plate). In this way, the beekeeper can easily examine the mite fall without disturbing the bee colony. The natural mite fall gives an indication of the degree of infestation in the bee colony and the success of the treatment can also be assessed with the mite fall after a treatment. The diagnosis of mite fall is very easy to do: one slides in the drawer and takes it out again a few days later, after which the fallen mites can be counted. Depending on the time of year and the occurrence of mites, the drawer is left under the bee colony for 3 to 10 days, shorter in summer and longer in winter. See later the chapters about varroa and their control.
- with the drop diagnosis too on the drawer, the strength and posture of the bee colony can also be seen and assessed.
- the mesh bottom is sufficient for ventilation when travelling. It makes the purchase of a relatively expensive travel front or other travel devices unnecessary. The preparations for travelling are completed in just a few steps: the flight hole is closed with a suitable foam strip, a tension cord goes around the box and you're done.

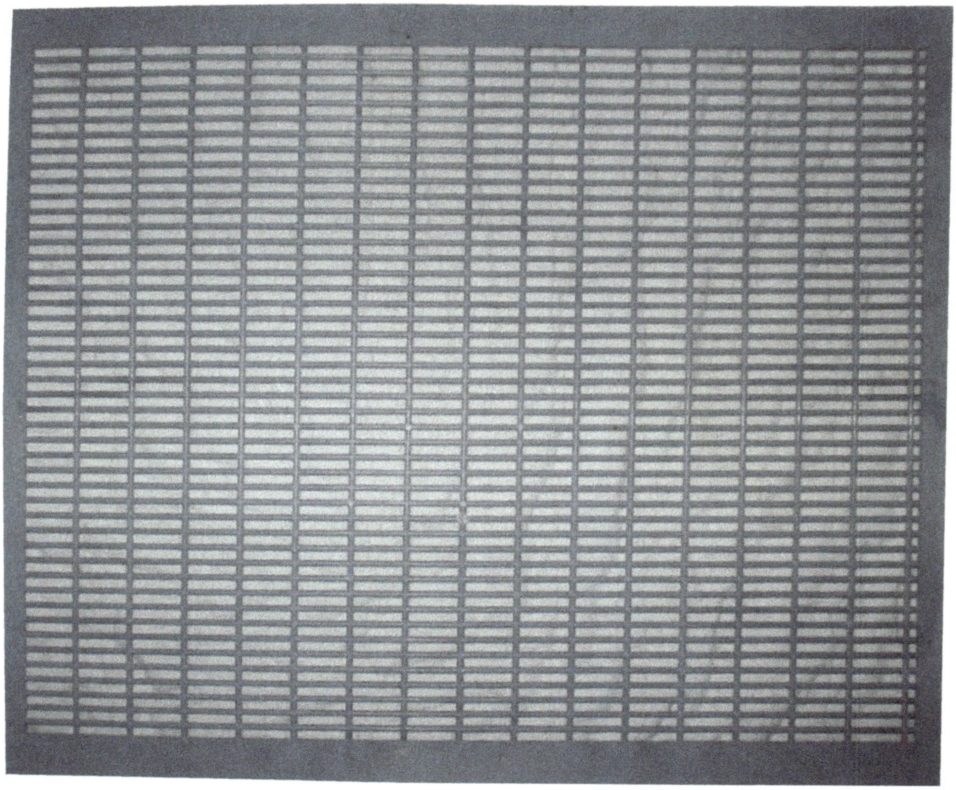

queen excluder made of plastic

Beekeeping with a queen excluder

Using a queen excluder, the honey chamber is separated from the brood area by the two boxes with (usually) ten frames each. The size of the openings of the queen excluder is measured in such a way that only the workers can move through the mesh. The queen and the drones are thus denied access to the honey box. The development and in particular the breeding activity of the bee colony is not influenced in this way. The remaining 20 frames are, if building frames are applied for drone cells, actually 19 or 18 frames respectively. If one or two frames are used for drone cell construction, there is enough room for the worker brood, pollen and food. A fully developed Zander frame contains about 6000 cells. If the queen lays 2,000 eggs every day for three weeks, the brood nest will contain 42,000 cells. For that number numerically seven Zander-frames are sufficient.

Theoretically, one can therefore limit the breeding space to one box. Of course, that becomes a bit too scary because the bee colonies also need space to store a supply of pollen (as bee bread) for the brood (almost half the space size for the brood) and, with good flow, the brood nest can be turned into a honey chamber. Interim storage of the nectar or honeydew carried in, leads to a reduction in breeding activity much sooner and more strongly than in a brood area with 20 frames. In any case, 20 brood frames are sufficient, even when beekeeping with a queen excluder.

The queen excluder leads to a clear separation of the honey space and the brood space, which considerably simplifies the care of the bee colonies:

- In the case of swarming colonies, only the 20 frames of the breeding area should be checked for swarm cells. If no queen excluder had been used, all frames of the brood area (30 frames) had also to be handled and examined.
- The honey harvest is significantly easier because no honey frame is occupied with brood. If no queen excluder had been used, the frames with worker brood or drone brood, even if they contain more than 2 kg of ripe honey, must remain in the colony until all that brood has hatched. One simply does not destroy brood.
- In addition, a bee escape (see also honey harvesting) can be used to prepare and remove the honey frames. "Overnight", so to speak, the honey frames then become free (of course only if there is no brood on those frames as should be) and can be removed the next day, box and all, and exchanged for a box with empty frames.
- Anyone who does not use a bee escape and removes the honey one frame at a time will also have an easier time if a queen excluder is used. He does not have to pay attention to the queen and can sweep the bees in front of the entrance, of course only if the bee colony is on or close to the bottom and not on a high scaffolding. In that case, the beekeeper must use a board or the inner lid to make the connection between the bottom and the entrance (as a "climbing board"). Only then is it guaranteed that all bees wiped before the entrance find their way back into the hive. If the bees are shaken or swept into the box when the frames are removed, some of them will be dealt with several times, namely bees that, after being wiped, immediately crawl to an adjacent frame and then for a second or third time, need to be swept. Then many bees can become aggressive and sting. If they land in front of the entrance when wiping, the risk of stinging is reduced.
- When using a queen excluder, species of honey can more easily be harvested. If a following flow needs to be used after a first blossom period, it is recommended to empty the honey chamber. If no queen excluder has been used, a lot of honey from the previous honey crop will remain on the frames together with (some) honey on the brood frames in the colony. This way than only a mixed honey can be harvested.

- Protecting the frames in storage against wax moths is easier to achieve if the bee colonies are provided with a queen excluder in the spring and summer. In any case, the frames in the honey chamber remain brood-free and can be stored after extracting without the risk of them being ruined by the voracious wax moth larvae. Of course, they should not be stored together with frames that have been incubated or still contain pollen stores.

- If only clear, unincubated frames are used in the honey room for honey production, unsealing can also be done with hot air. That's easier than with an unsealing fork. But the unincubated frames are less stable than incubated ones. When extracting, it can easily lead to honeycomb breakage. Therefore, the extracting device should not be run at full power.

- The breeding schedule already pays off in the spring when the colonies are placed more widely. You can assess the condition and development of your bee colonies, as well as the flow progress, much better if the honey area is separated from the brood area on two boxes by a queen excluder. The stronger colonies use the honey box sooner than the weaker colonies and they build also the foundations faster.

- If the flowering period is interrupted by a period of bad weather, the honeycombs are abandoned again unless they are already filled with honey. This upward and downward movement can be observed particularly clearly in changing weather on the foundations that are hung on the sides. When nectar is brought in, it is occupied and expanded by the construction bees. In cool weather, the full or only partially extended foundations are completely free of additional space. The construction bees are needed at the bottom to keep the brood warm and they are only go back up when further construction is worthwhile. Construction activity will only resume if something is brought in from outside. If no queen excluder has been placed, the colonies will in any case, even in bad weather, move upwards and incubate the core of the container that was placed when the cage was moved to a larger size. Even if a massive flow subsequently occurs, this brood cannot be immediately pushed down. The colony stretches outwards rather than move downwards again. Optically, these colonies are more likely to need a fourth box than colonies that are kept under a queen excluder. Of course only optically, their lower brood box usually contains little or no brood. With moderate flow, the colonies without a queen excluder also incubate the fourth box, which is then actually not necessary.

A view on the queen excluder made of metal seen from a honey box after removing the honey frames. (Photo Dr. G. Liebig)

Feeding honey bees

Feeding the bees is necessary for wintering. Feeding is absolutely necessary once the honey has been harvested. Feeding is necessary even if the bees have produced little or no honey. This feeding is best provided from above or from the side in frame feeders. When young colonies are formed, they also need to be fed. This is best done from the bottom or side in a plastic bowl. In the winter when additional feeding is required, the bees are fed from below in a plastic bowl with sugar water or syrup that is placed directly under the bee cluster. This is more animal-friendly than a slab of fondant on top of the combs because no water needs to be used to liquefy the food.

Household sugar suffices as food. It is dissolved in water, preferably in a ratio of 3:2 (3 parts sugar to 2 parts water) and then given to the bee colonies in a feeder. Instead of household sugar dissolved in water, syrup made from sugar beet or grain starch can also be used. This is available at specialist beekeeping stores.

Feeding food for winter "from above" or "from the side" is preferable to feeding "from below". This is more hygienic and does not provoke robbery as easily as feeding "from below". A bucket, stacking box, saucer or bowl can be used for top feeding. It is then placed on top of the colony in an empty box. Board feeders can also be used, which are hung in an empty box above it on the side or otherwise in the top box near the wall. In the latter case, two frames must first be removed.

Swimming aid is given on the liquid food so that the bees do not drown. Suitable for this are untreated straw, untreated wood shavings, bottle corks, grass, foliage and short-cut twigs from shrubs, deciduous trees or pine trees.

In feeders with smooth walls, the bees also need a place to climb out.

The empty box is tightly covered with the inner lid and the metal roof during feeding. When refilling, there is contact with the bee colony. Nevertheless, one does not need a hood or veil or smoke.

Feeding bucket, plastic dish or stacking box with an empty box is considerably cheaper than an extra feeding trough and they are also easier to clean. They also require less space in storage. The empty box is also used before and after feeding for treatments against varroa in late summer and autumn. The things that are needed when using formic acid or oxalic acid are placed in the box and on top of the colony.

Feeding "from above". The bowl stands in an empty container placed above it. The foil under the tray is slightly pulled away or is placed diagonally on the tray occupied by the colony, so that the bees have access to the food from the front (or from the back) through a slit. The foil prevents the dish from being built on. The laid grass serves as a swimming aid. Climbing aids must also be provided for smooth walls.

The choice of frames

The frame size has been a topic of discussion for many generations of beekeepers. Even now there is a lot of arguing about which size frame is best. In the argumentation, one only needs to take operational technical problems into account, because it does not matter for the development of the bee colonies whether there are many small or few large frames available in the breeding area. The bees adapt flexibly to such a situation. Large frames are less convenient to handle for the beekeeper than small ones and small frames require more handling than large ones. The optimum lies thus somewhere in the middle... When choosing the frame size and the type of box accordingly the following considerations should be taken into account:

- The same frame size should be used in the breeding box and in the honey box. That way the building performance of the bee colonies in the honey box (= expanding the foundations) during the flow (from April to July) can be used for the later renewal of frames in the brood box. This exchange of frames (old against new) must be carried out every year. It prevents the occurrence of diseases! Regular comb renewal is an essential part of late summer care and is carried out box by box. Then the advantage is taken of the fact that with the natural contraction of breeding activity from the end of July, the shrinking brood nest moves upwards and in the two-shared breeding space. So, the lower brood box (with the oldest combs) becomes brood-free. After the brood has fledged (after mid-August) this box is removed completely. In its place - on the mesh bottom box - comes the top brood box with the combs that are one year younger and on top of that the honey box with the still virgin combs. Anyone who uses a smaller frame size in the honey box than in the breeding boxes has to take additional and time-consuming care about comb renewal in the breeding area. Old brood frames must be replaced one by one with foundations. In order for them to become brood-free, they must be placed aside to the wall of the box before the exchange. There the queen rarely fills them with eggs. After at least 21 days (and for the drones 24 days), all brood has hatched and the frame can be removed and replaced with a foundation. This foundation is only built up with wax if honey needs to be stored in it. Large or high brood frames often still contain food and pollen at the top. This stock is lost when the old combs are removed and melted. In breeding areas with two boxes, the food and pollen supplies in the combs are concentrated in the upper brood box that stays and that way it remains with the supplies in the colony.

- The breeding area should be divided in two. This makes swarm control considerably easier. During the tilt control, the breeding space is pulled apart by tilting the upper breeding box. A glance into the newly opened breeding nest is sufficient to judge whether the colony is in swarming mood or not. If necessary, a flashlight can be used for a better look. In a breeding area on one box with large frames, some if not all frames must be opened to be able to rule out with certainty that a colony is in a swarming mood. Before this, the honey box must be removed, even if it is still very light. In swarming colonies, checking (for swarm cells) of a non-shared breeding space due to the smaller number of breeding frames apparently makes less work, but has the disadvantage that no tilt control is possible. This becomes clear with a calculation example: With normal population development and beekeeping, three out of ten colonies become in swarming mood. In a hive system with one breeding box, in principle, also every swarm check that must be carried out week by week from the end of April to the end of June. During every swarm control all breeding frames of all ten colonies must be lifted and checked in order to determine which colonies are in swarming mood and which are not. In swarm-like colonies, this checking is connected with breaking the swarm cells. In a hive system with two breeding boxes, opening all frames is only necessary for three out of ten colonies, because a simple tilting control is sufficient to recognise swarming tendencies.

- During the forest flow period the colonies shrink so that towards the end of the flow period the majority of the pine honey brought in is stored in the breeding area. It should not remain there because it is not suitable as winter food and is also harmful. Removing them from a breeding space is more complicated in an undivided breeding space. In addition to the forest honey, there is often brood on the large windows, which cannot be extracted into every extracting device.
- The varroa control concept of "split and treat" is also considerably easier to implement if the same frame size is used in the honey box

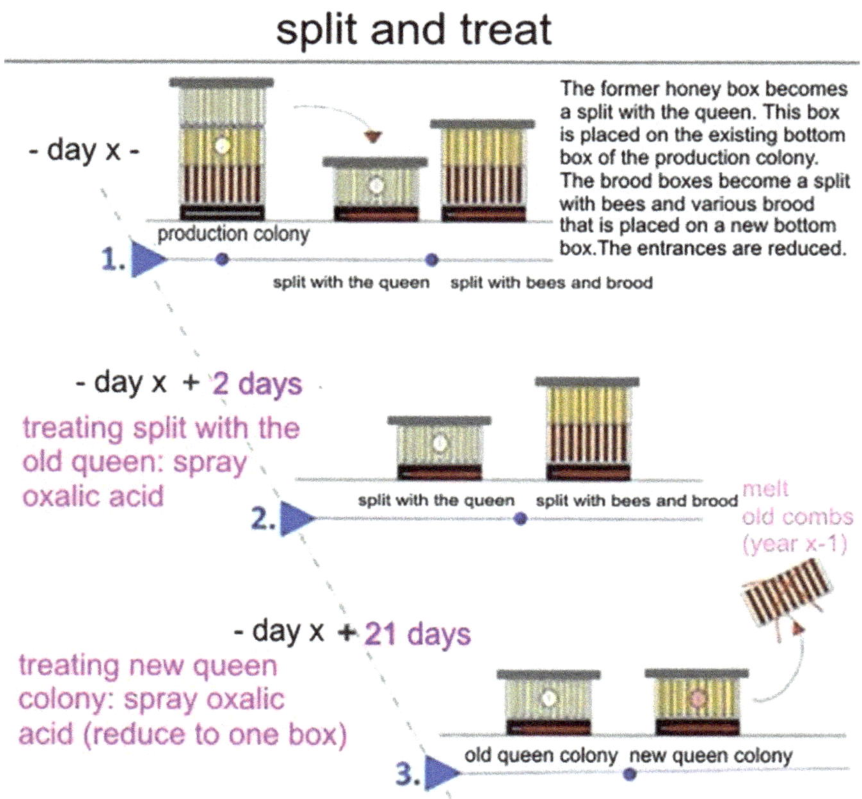

after the original figures of Dr. O. Boecking, Bieneninstitut Celle

and the brood box and the colonies are kept in a brood box divided into two with a queen excluder and a honey box on top until the colony is divided. The honey box then becomes the split with the queen (with the "old" queen). The new brood nest is formed on the clear combs therein. The queenless remaining colony (= the breeding colony) is reduced in size after all the sealed brood has emerged. So, after 21 days. The bottom brood box with the older dark combs is removed.
Both parts of the colony can then be treated with oxalic acid against the varroa mite in the brood-free state, the split hive with the queen two days after the colony division, the 'breeding colony' three weeks later.
It is assumed that all boxes of the hive are of the same size. So the super has the same format, is the same size, as the brood boxes.

Dadant, Langstroth, Zander and hives with shorter "ear" frames

Based on the considerations described above, the purchase of a Dadant hive is not recommended. Its breeding area includes 12 large Dadant frames and the honey area is equipped with ten clearly smaller thick frames.

In general, the use of thick combs in the honey room should be discouraged as the honey in thicker frames is usually stored with a higher water content than in normal "thin" combs.

At Langstroth, Zander and comparative format boxes, the colonies are best kept in a breeding area divided into two. The same size frame is used in the honey room as in the breeding room. The Langstroth frame is the most widely used worldwide. The Zander format is slightly smaller The disadvantage compared to the Langstroth window was the thinner top bar, but that has nowadays been overcome by adaptation (See also picture on page 19).

The colonies create comb on top of the frames or between the boxes, which interferes with working in the hives. The latest Zander frame variant is now increasingly being used with the modified or thicker top bar to solve this problem. The advantage of the Zander frames is that they have longer handles. This makes the frames easier to handle. The so-called modified Zander frames, which have a top bar with the thickness of a Langstroth frame, are ideal.

Some traditional smaller format frames have less surface area than a Zander frame. Ten Zander frames correspond to approximately twelve of such frames. Ultimately, working with these size means 20% more work: when manufacturing and wiring the frames, when casting wax and installing foundations, when lifting the frames, when harvesting and extracting the honeycombs.

The bees don't care what size frame they are on. It does matter to the beekeeper. It's obvious by now that I favour the Zander format.

You always have a good grip on a frame with long ears, as can be seen in this photo when lifting a construction frame [drone brood frame] and then shaking off the bees that are still on it.

(Photo: Dr. G. Liebig)

How to make a foundatation frame with wire and wax

At least in the UK, wax plates are also offered for foundations that are reinforced with wire. There are many nstructions on the internet how to attach these plates with iron wire in a frame. If you want to save the money that is involved, it is best to make frames with steel wire yourself. Espcially if wax is also obtained from the old combs or drone cells building frames that can be pressed into plates. Making a frame with steel wire is very simple.	
The tools needed:	a hammer, a (combination) pliers, a measuring tape/roll, a wire cutter, - an awl-, a transformer (> 24 V) or a car battery.
Materials (Zander): If you have a hive with different dimensions than a Zander hive, different frame dimension may apply.	Stainless steel wire (ϴ 0,5 mm), preferably stainless nails or galvanized nails (1.2 x 20 mm), wood glue Frame parts: top slat, frame sides, bottom lath -metal (brass, copper) eyelets- Dimensions: (all dimensions in mm) Sides: 477 x 220 - 4 x perforated, 10 thick, 35 Hoffmann sections. Top slat: 477 long (with 8 mm groove), 20 thick, 27 wide
	Bottom lath: 420 long (with 3 mm groove), 10 thick and 22 wide (Ears: 28.5 long, 10 thick, 22 wide Wax plate(s)

All the things together you need to construct a wired frame from prefabricated frameparts in Zander format.

Make sure the sides of the frames have openings that can be used for metal (brass) eyelets.

To assemble the frame you will need little nails (1.2 x 20 mm is best) and wood glue.

Traditionally, no wood glue is used, but the falling apart of a bee-infested frame when you are working in the is no fun.

Put some glue in the fork of the frame side. Press the (top) slat in the fork of the frame. The top slat in the picture is the modified version.

Hammer a nail into each side of the frame side to further secure the top and the bottom slats. On one side, hammer the nails in at the top and bottom for about half. These will then form the attachment points for the iron (steel) wire.

The two attachment points for the iron (steel) wire.

(A modified frame is a frame with an extra thick top slat. If later the foil fits well on the top of the frames, the honey bees will not build comb on top of the frames so quickly)

A simple frame like this -but with all nails hammerd in- can be used as a building frame for drone cells.

Now the frame is ready to be wired.

For a Zander frame approximately 2 m of steel wire is required to provide the frame with wire.

Wrap the wire around the nail at the top stat a few times and in such a way that the wire remains attached to that nail.

Using the combination pliers (or other pliers that may suit), pull the wire taut and then insert the wire on the outside into the nearest perforation of the frame side.

Pull the wire through all the openings and tighten the wire on each frame in- and outside with the pliers.

Use the pliers to twist the end of the wire tightly and securely around the nail at the bottom slat.

Cut away the loose wire ends and hammer the nails all the way in.

Now we have a wired frame that is ready to be fitted with a wax plate.

Slide the wax plate into the slot of the top slat. If there is no slot, place the plate on the wired frame with the top connected to the top slat.

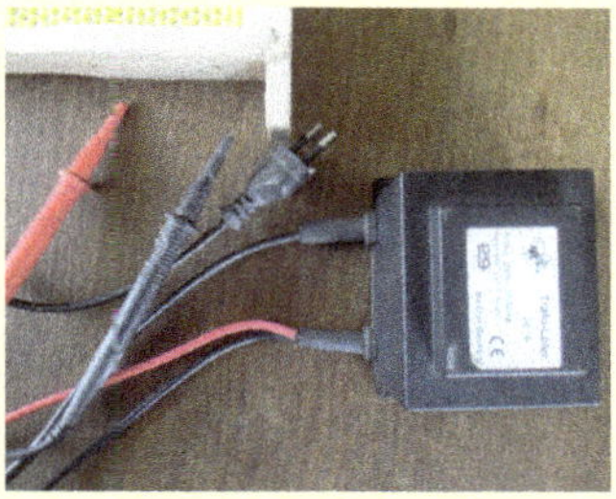

For fixing the wax plate to the wire of the frame by slightly heating the wire a car battery can be used.

With a soldering transformer the process goes faster. Cinnect the poles of the transformer (provided with electricity) or the battery shortly to the wire of the frame at the two nails that secure the wire. Make sure the wire does not get too hot and the wire cuts the wax plate into pieces.

When you start to see the wire through the wax plate already a little, remove the poles from the wire immediately.

Now the wax plate is attached and the foundation frame is ready. Hang the foundation frame (for example in an empty brood box) so that it is not damaged and the wire can cool down quietly.

Ten frames in a box

Depending on the frame size and the hive system, a box can accommodate between eight and twelve frames. In the breeding area of so-called trough boxes, even up to 20 pieces. Hives with Zander format frames are available in constructions with 8, 9, 10 and 12 frames per box. A box with eight or nine full honeycombs is easier to carry than one with ten, but during the honey harvest everyone is free to fill the transport boxes with only as many frames as can be carried. The design of the breeding area or the wintering area is of greater importance. In both hives (with ten frames) the breeding space can develop in width and moves upwards earlier in late summer than in narrow hives (with eight or nine frames). The bottom brood box will then become brood-free sooner. For young colonies, a box with ten frames is usually sufficient for overwintering. With eight or nine frames, adequate supply becomes a problem.

For changing frames with different dimensions, simple aids are enough. In the pictures above, a bar with a small slat is enough to convert from small to larger or vice versa. This is one example to point to the solution you might need.

Cold construction

In the so-called warm construction the frames are perpendicular to the entrance, in cold construction they are arranged towards the entrance. It does not matter for the development of the bee colonies whether they are in cold or warm constructions. But the bees can more easily regulate the formic acid in the hive air in cold construction if it is used in late summer to combat the varroa mite. The beekeeper can then safely dose so high that proper functioning is guaranteed without endangering bee health. The tilt control also makes it easier to view the comb seams in cold construction.

The inner lid

The single-bin stacking hive has two lids, an inner and an outer lid. The inner lid consists of a simple wooden framework that is nailed or stapled to a three cm thick (pressed) wooden plate. Two short slats are attached under the ends in both corners of the ends in such a way that the inner lid fits snugly on the container and cannot shift when placed flat on it. The inner lid can also be placed the other way around ("wrongly"), with the concave side downwards.

In this way, the colony can also be tightly covered when a slab of feed dough is placed on top for emergency feeding.

Also when placing a queen in the colony, the queen cage can be placed on the colony and does not have to be hung in a comb alley. One can therefore clearly observe the behaviour of the bees around the cage.

Finally, but seldom applied, if food is offered in a board feeder hung on the edge, a suitably cut twig can be placed perpendicular to the frames, which makes it easier for the bees to access the board feeder.

Foil

We usually place a plastic foil between the inner lid and the bee colony. This prevents the inner lid from sticking to the combs and makes it considerably easier to open the hive. Placing a foil has no influence on the water content of the honey, for that matter.

Each hive has a metal roof so that no further covering of the colony is required all year round.

An ideal inner lid: flat at the bottom, with a "raise" at the top. During emergency feeding, the piece of food dough placed on the colony is nevertheless covered tightly with the inner lid.

Photo: Dr. G. Liebig

Protection against mice in wintertime

During the winter half of the year, the bees that are in a cluster must be protected against mice that want to enter the hive. The entrance must therefore be reduced to a height of 7mm so that even the smallest shrews are denied access. Mice that enter the hive ruin the comb building, eat bees and worry the bee colony. Weaker colonies can become victims of mice.

To make smoke, life can be made easy if a gas burner with a piezo element is used to make the straw pellets glow and produce smoke. Straw pellets used as bedding for rabbit hutches, for example, are excellent for use in your smoker.

A mouse slat or mouse mesh can serve as protection against invading mice. A mouse slat reduces the flight gap on all sides to a certain width and can therefore also be used if there is a risk of robbing. Of course, a mouse slat should only be used when the colony is sitting in the middle!

Small entrances in winter can have catastrophic consequences.

Particularly in the case of prolonged cold, so much dead bee fall on the ground that the entrance becomes blocked. When the weather becomes milder and the bees that want to go outside cannot leave the hive, the colony will perish. The bees panic, eventually become exhausted and die. The brood is not cared for further and dies too. Small entrances should therefore be checked again and again in winter for accumulating dead bee fall. Clogged flight gaps must be cleared before flying weather occurs again.

Using mouse mesh for the entrance that has not been reduced in size will provide adequate aeration, even if a lot of dead bee waste should accumulate on the bottom.

A mouse mesh applied in wintertime. As soon as it gets cold, the strip of foam used to reduce the entrance can be removed.

The bees have some difficulty getting through the mesh. This applies in particular to the workers that return home with pollen. As they squeeze through the mesh, the pollen baskets are occasionally wiped off and the pollen is lost.

The cleaner bees usually try in vain to remove the dead bees from the hive. The dead bee fall usually only accumulates behind the mesh when the colonies start the major cleaning during flying weather in spring or already in the winter time. Nevertheless, the mouse mesh is preferred because it makes monitoring unnecessary in winter. The mouse mesh will be removed in March at the earliest when it becomes clear that flying weather will prevail every day or almost every day.

Painting the hive

The beehive is, if desired, but not necessary, only provided with a protective layer on the outside. Almost any wood stain that is harmless to bees is suitable for this purpose. The colour is according to the beekeeper's taste and wishes, but it looks better if it is adapted to the environment.

A painted wooden hive makes little difference in terms of durability compared to an unpainted hive. As Dr. Liebig always says: "A painted beehive lasts 30 winters, an unpainted one 30 summers..."

Honey bee biology introduction

Numbers

A niche is a small defined or demarcated part of factors or conditions under which a certain specialisation can thrive. This is a general, open definition applicable to several scientific disciplines such as Economics and Biology.

Also: an ecological niche is the place that a species or population occupies in an ecosystem, or an n-dimensional hypervolume within which the species can maintain a stable and viable population. A niche is the role that an organism plays in an ecosystem.

Each niche is the result of ongoing natural selection. Due to competition and changing circumstances in the long term, life continues in accordance with the biotic and abiotic factors available in the biotope. This is an important notion. It is the meaning of the survival of the fittest adage and marks the fundamental idea of evolution: adaptation per generation to the ever-changing life situation.

Thus, as an abstract metaphor for thinking, time can be considered the 'niche' of humanity, of the species "human beings". The primary niche of honey bees, so thought and said, is in this view 'number'. Although honey bees share this niche with other organisms (with which they -therefore- have no food competition...), this characterisation as such is not exclusively unique or even entirely correct, but it does illustrate a basic understanding of honey bees and how a honey bee colony thrives. (It is tempting to speak of "functioning" rather than "thriving". However, it is easier to understand conceptually that each member of the honey bee colony has a specific role. In workers there is even a certain plasticity in which role to take at any given time. A 'dynamic' that seems a bit more difficult to understand with almost mathematically thought or fixed functions.)

A queen alone is helpless, a drone (a male bee) is as well, and a worker alone cannot cause much in the end. But together, a queen, a few drones and a lot of workers, all cold-blooded insects on their own, that is something different, that is even almost like a warm-blooded organism.

Also notable is the analogy between bees and their food sources in terms of reproductive strategies: drones and pollen, queen and pistil.

Most people are already familiar with honey bees. Animals that produce honey, that will sting if you disturb their nest and that have a queen that lays eggs. Sometimes in secondary education this knowledge is in the biology lessons augmented with the main aspects of the honey bee dance. An amazing way to communicate in the dark of the hive the place where nectar is available while taking in account as well that the position of the sun is changing because of the earth rotation in the course of time. Sometimes also the division of labour (which will be discussed in more detail later) and the importance of pollination by the honey bees is mentioned.

A honey bee colony is fascinating for many, especially for those who have remained aware of the enchantment of nature. The experienced beekeeper realises the many miracles of a honey bee colony and the astonishing properties of honey bees, ofte also due to questions from curious visitors.

An important condition for successful honey bee care is a fundamental knowledge of honey bee biology. This does not mean that only people who have thoroughly studied scientific books on honey bee biology can keep honey bees. But you must first familiarise yourself with a certain minimum of knowledge about honey bee biology. Everything else will then take care of itself. Anyone who becomes fascinated automatically feels the need to learn more about honey bees. Also this is the reason to restart this honey bee introduction simply.

What should a beginner know? Personal knowledge is always limited and can always be increased. The more you know, the easier it is to deal with honey bees. On the other hand, there is no need to study the biology of honey bees every day late into the night.

The cognitive capacities of a honey bee

It is almost unimaginable what already a single honey bee can do. She can navigate and orient herself as a home centered insect within a radius of five - eight km from her hive. She can also remember the location, and the type of flower by shape, odour and colour and communicate all this when she has returned in her hive. And this all with a brain of the size of 1 mm^3 (with about 970,000 neurons). For comparison: our brain is about 1 350 000 mm^3) (ca. 10 tennis balls) and we use about 90-100 billion neurons.

The honey bee forager is an animal that is programmed to learn and to memorise multiple features constantly in the field. She visists the same flower species as long as it provides nectar or pollen. She does not switch from one flower species to another (as 'opportunistic' bumble bees do). This is called flower constancy and by this we can enjoy various types of honey according to the type of flower that has been visted by the honey bees as they work in a chain as well.

Karl Ritter von Frisch, picture constructed by the author, original photo source is unknown (internet)

The communiction capacities of honey bees are amazing and fascinating. Besides communication with pheromones (chemical signals) and trophallaxis (direct transfer of food and fluids from one individual to another) honey bees communicate also where to find food or a nesting place. In the dark of the hive they perform a waggle dance. A very special form of communication discovered by Karl Ritter von Frisch. This is a performance done by forager bees that have returned to the hive and start to communicate the location where they have found the food they brought with them. The forager walks in a figure-eight pattern on the comb and shaking in the middle of that pattern with their abdomen. This pattern is used for food sources more than 100 m away from the hive. Within a radius of about 100 m the forager will dance round shaped pattern. The direction the forager waggles corresponds with the direction of the location of the food source. The forager uses the top of the hive to indicate the position of the sun. The sun is the point of reference. The angle of the food source location with the sun is the same angle the waggle dance will be be made. They also take into account the changed position of the sun during the period when they have left the food source. The intensity of the waggling indicates the distance of the food source from the hive. When the colony is about to swarm, scout bees will also perform a waggle dance to communicate a location of a new nest site.

When the forager or a scout bee is convinced she has discovered a richer or better location she competitively frustates her sister or half-sister's waggle dance by bouncing off her with her own waggle dance.

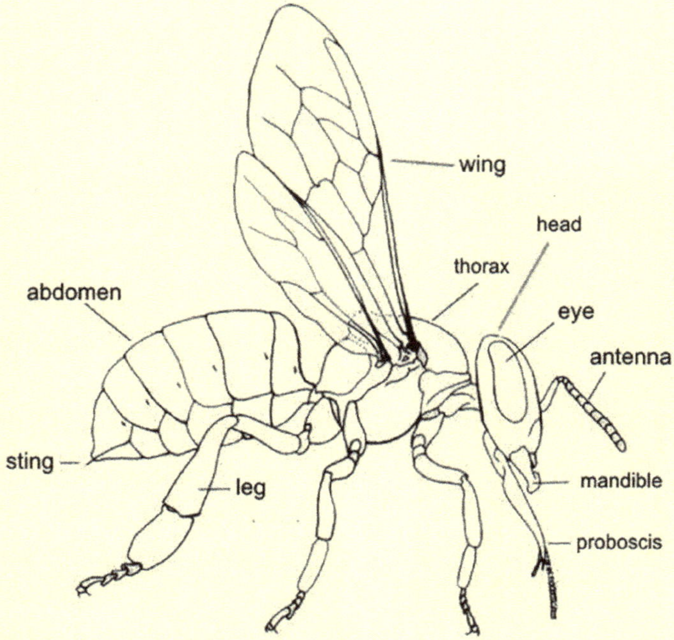

Simplified honey bee morpholgy. When reading about honey bees it is of some importance to be familiar with the names of the main body parts of honey bees. The two antennae at the front are flexible appendages used for feeling and smelling. Honey bees have multiple mouthparts. The mandibles are used for grabbing food, working with wax, and biting. The proscobis is a retractable suckuing mouthpart consisting of multiple tubes, used for drinking water, nectar and honey. The thorax is the middle body segment of the honey bee to which the wings and the legs are attached. The abdomen is the back end of the body that contains important organs to sustain functional life.

The honey bees can do so because they can learn and memorise in a very efficient way the features of the flowers which they are exploring at the time. They learn the colour, they learn the odours, they learn the shapes, they learn the position in space. So, indeed honey bees can see colours, as already was found by Karl von Frisch. However they cannot see 'red', but in stead of this they can see 'ultra violet' and they can make use of polarised light (which is as very usefull as well if the sun is not shining brightly). By all these opportunities of learning it is possible for them to return to the same spot, to the same flowers as long as these provide the food they are looking for.

Besides elemental forms of learning, associative forms of learning (for instance: a honey bee, at the blue dish she finds sugar water, at the yellow dish there is nothing) they manage as well higher order forms of learning. Honey bees are capable of conceptual learning, learning about relations, relational learning. They can learn the relational concept of sameness independently of the physical nature of the stimuli trained. They can transfer relations for instance between colours and odours or with another or new type of stimulus than the one they were initial trained with. This is even the more astonishing as the bees can learn the relational concept of sameness much quicker than other animals, as for instance a chimpansee. Also they can learn the contrary, the concept of difference. Also they can use the concept of 'nothing'.

They can learn several concepts as for instance spatial concepts like 'above' and 'under' and 'to the right of''and 'to the left of'. Honey bees can also master two rules simultaeously and they can learn numerical concepts (they can count from one up to a limit of five) and also honey bees have a form of awareness.

And never forget: honey bees can distinguish and recognise faces.

All this knowledge dates from the recent past. It is fascinating and if you have the opportunity to learn more about it, don't hesitate to do so. More research in these fields is underway as for instance the research to the effects of having a brain divided in two hemispheres as other animals (including human beings) have as well. And it is really tempting to believe that the honey bee brain is comparable more efficient than the brains of other animals.

Bees and honey bees

Our honey bees are usually referred to simply as bees. This ignores the fact that in Centra Europe alone, in addition to honey bees, there are hundreds of other bee species, including bumblebees, for example. Honey bees and bumblebees are social bees, they live in a colony, the honey bees constantly, the bumblebees only in the summer. The overwintering young female bumblebees start in the spring with a nest that grows into several dozen or a hundred individuals until the summer or autumn and dies off after the arrival of the sexual animals. The other bee species are hermits or solitary bees and are usually included -together with bumblebees- under the concept of wild bees. This concept expresses the fact that solitary bees and bumblebees are not cared for by humans and live in the wild. This is, for that matter, despite the fact that more and more bumblebees are being used for pollination in greenhouses. There are also wild honey bees in the forests. These are swarms that have sought the distance and thus escaped beekeeper care. Every year many swarms have succeeded.

All bees are good pollinators. The statement "No fruits without bees" is certainly largely valid if it refers to all bee species. Only in relation to honey bees would that statement be greatly exaggerated as is estimated that only the half of all flowers are pollinated by honey bees. That is nevertheless a great achievement.

Caught! A worker bee feeds the queen larva.

Photo: M.de Gelder

The three honey bee castes

Honey bees live in a colony. In every honey bee colony there are three sections or manifestations of life: the queen, the worker and the drone. Drones (males) are usually only found in summer. In winter, a honey bee colony usually consists of a queen and her workers (daughters). The number of workers varies during the year between several thousand during the winter and many thousands in the summer. Its environment (location, weather conditions [climate], flow [available flowers] controls the development of a bee colony. A special phase in population development is the swarming time in May and June (in our part of Europe).

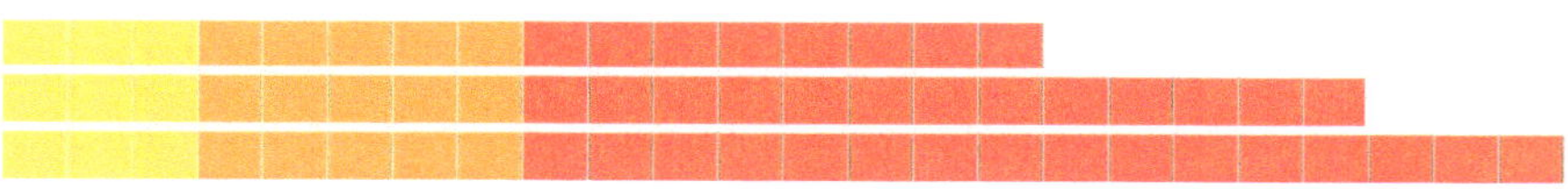

From egg to honey bee in days (queen, worker, drone)

Queen (from egg to insect, 16 days)

Three days egg, five days larva and eight days pupa ››› three, five, eight and the queen is made

Worker (from egg to insect, 21 days)

Three days egg, five days larva and thirteen days pupa; 21 days in total

Drone (from egg to insect, 24 days

Three days egg, five days larva and sixteen days pupa; 24 days in total

When caring for his colonies, the beekeeper pays attention to population development and tries to take into account the influences of environmental factors. Workers, queens and drones each contribute equally to the functioning and survival of a bee colony. Most of the work that needs to be done is done by the workers, but without the queen and drones the honey bee as a species could not exist. The versatility of tasks that the workers perform or can perform in and for the colony is proof of the high degree of development of honey bees and the special skills of the workers. They are truly all-rounders. The worker honey bees perform their tasks usually at a specific age, also mainly due to their hormone composition at a certain age. Nevertheless they can go back to tasks in earlier age if this is needed so.

The honey bee colony is, like no other animal of economic use, very dependent on its environment. It can only maintain itself in constantly changing circumstances if the understanding between the many individuals functions without friction. That is therefore usually the case.

Day by day, day and night, in all weather, in good or bad flow, in spring, in summer, in autumn and in winter.

Although there is a hypothesis that in colonies with a hybrid origin of different honey bee subspecies some irritation between the half sisters is caused by differences in 'dialect'.

Worker tasks / activities	***Worker age ****
Cell Preparation:	2-16 days old, mean of 8 days
Capping Cells:	3-10 days old, mean of 8 days
Feeding Larvae:	3-15 days old, mean of 10 days
Queen Tending:	3-14 days old, mean of 11 days
Honey Production:	10-20 days old, mean of 13 days
Cleaning Debris:	11-15 days old, mean of 14 days
Pollen Handling:	10-19 days old, mean of 15 days
Comb Building:	12-13 days old, mean of 16 days
Cell Cleaning (again):	11-25 days old, mean of 13 days
Ventilating the Nest:	2-25 days old. mean of 19 days
Hive Guarding:	15-28 days old, mean of 19 days
Orientation Flights:	17-27 days old, mean of 23 days
Foraging:	18-28 days old, mean of 24 days
Assorted Tasks:	
Dancing	
Bearding	
Thermoregulation	
Egg Laying (Drone Mothers)	
Patrolling	
Defence Response	
Resting	
Robbing	
Grooming	
Trophallaxis	
Swarming	

* During their shorter lifespan in summer (often about just a fortnight) the list of task ages is than also more dense. Resting is also a task. It takes normally about 50% of their time.

Yes, it all starts with an egg

Every bee, whether it is a queen or a worker or a drone, develops from an egg that the queen lays in either a worker brood cell, a drone brood cell or a swarm cell. These three cell types differ in size. Drone (brood) cells are significantly larger than worker (brood) cells, but have a six-sided shape in common with them. Newly built worker (brood) cells have a diameter of 5.37 mm. The diameter of drone (brood) cells is 6.91mm.

When the worker cells are used for brood, they are between 10 and 12 mm deep. The distance between two brood frames is so great that the worker bees that care for brood do not get in each other's way and both surfaces can be occupied with bees.

The application of this so-called bee space (a gap between 6,4 and 9,5 mm or 1/4 and 3/8 inch) everywhere in the hive is invented by Lorenzo Lorraine Langstroth in 1851 and patented in 1852. The idea led to the development and the use of bee hives with removable frames as we still use tem to this day (the Langstroth hive is in fact the oldest bee hive with removable frames, removable frames that are not glued by the bees with propolis to the bottom, the walls or the deck board / inner lid).

If the queen is on a comb densely populated with workers, it can easily be overlooked despite her size. Therefore it is recommended to mark the queen.

Photo Dr. G. Liebig

If the cells are used for honey storage, the bees can extend them. The space between the combs then becomes smaller and only offers room for one bee. When the honey cells have been emptied and are to be used for brood again, the cell walls are shortened again and removed (demolished).

The hexagonal cell shape means that there is no empty space between the cells where dirt can accumulate. That makes cleaning easier for the workers. Moreover, the available space is optimally used.

When drone cells and worker cells are built next to each other, there are irregular transition cells between those cells of different sizes. Irregular cell shapes can also be found at the edge of the comb and especially where the comb is attached to the frame.

A comb has two sides. The cells are built on the foundation on both sides. The walls of the cells on one side are not opposite the cells on the other side, but half a cell's width away. The walls of three neighboring cells meet in the middle of the opposite cell. This construction bond effectively increases the comb stability.

Worker and drone cells give a comb its typical appearance. They are not level. Their long axis is oriented approximately 5° upwards. Play cells (often arriving swarm cells, queen cells) are generally built up at the bottom of the comb wall. They are not hexagonal but round. And they are not pointed upwards either. Their opening is usually located perpendicularly downwards. However, when the queen larvae grow up in it, they do not fall down. The food juice (royal jelly) with which they are so richly supplied also sticks to the wall. The queen larvae do not swim in it. They can't even swim. If the queen cells are placed upside down, the larvae drown!

During the development of the queen larvae, the play cells are expanded into queen cells and are thereby extended by 20 to 25 mm.

Fertilised or unfertilised

Workers and queens develop from fertilised eggs and drones from unfertilised eggs. Technically speaking, drones are haploid and workers and queens are diploid. Fertilisation takes place during egg deposition. The queen decides whether she wants to lay a fertilised egg or an unfertilised one. Before laying an egg, she inspects the cell. She can feel by the size and shape what kind of egg should be laid inside. The eggs laid in worker cells or play cells are fertilised with released sperm cells by the queen when laid. Eggs laid in drone cells have not received a sperm and remain unfertilised.

So, drones have no father, only a grandfather. This means indeed that eggs laid by workers are unfertilised and can grow, if not removed as happens in regular colonies, into drones capable to fertilise a virgin queen during the mating flight.

The eggs are attached to the "cell floor" with the pointed end. They stand there like inverted pencils (markers). That's why it's called sometimes: the queen 'pollinates' the cells when she lays eggs.

In worker and drone cells, the "cell floor" refers to the perpendicular rear wall.

The eggs laid there therefore do not stand upright in the cell, but stick almost horizontally and tilted slightly downwards to the back wall. This egg tilt is either perpendicular to downwards or at an angle of 30° or 60° to the right or to the left. The direction of the tilt is determined by the queen's behaviour during laying.

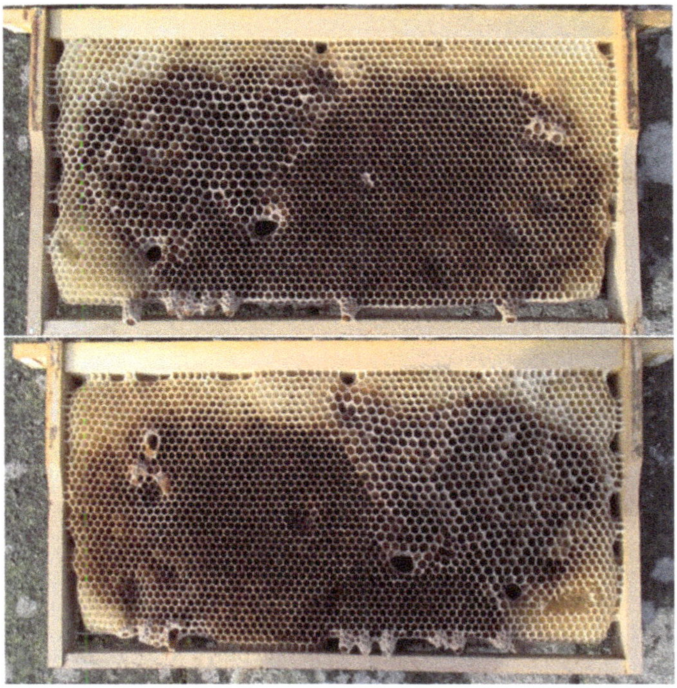

Two sides of a building frame with former worker and lager drone brood cells and some swarm cells on the bottom of the comb. There are even queen cells to see. The white sealed cells at the top edge of the lower frame side in this picture are filled with honey. It is clear that this beekeeper was a bit too easy.

In principle, the queen moves from top to bottom when searching for empty cells on a comb. To lay an egg, the queen inserts her abdomen into a selected and previously inspected cell. After the egg has been laid, she withdraws her abdomen at the same time as inspecting the next cell, using her antennae to check whether that cell is ready to lay an egg. That is never one of the three neighbouring cells, but always one of the adjacent cells. Five cells qualify for this, one of which lies perpendicularly below the cell just 'penned', two of which lie at an angle of 30° to the left or right of the cell perpendicularly below it or two that lie at an angle of 60° to the right or left of the cell perpendicularly below it. The newly laid egg tilts toward the cell that the queen inspects next. This behaviour has consequences. To completely and seamlessly 'pollinate' a comb, the queen must pass through that comb from top to bottom several times. And all this happens in the pitch dark of the hive. Egg laying involves impressive sensory feats.

In the queen cells, which are made for swarming or for a "silent queen exchange", the cell lid (wax lid) forms the "cell floor". After egg laying, the egg is not on the bottom, but hangs outright from the ceiling. The feed juice is also deposited there. The larvae that have hatched from the egg and are fed with royal jelly stick to the food juice and are therefore also on the ceiling.

When the queen's sperm supply is depleted, she can only lay unfertilised eggs. Drones then develop from the eggs laid in the worker cells, the hump brood. Such queens and their colonies are called drone layers. They are doomed if the beekeeper does not intervene.

During successful copulations during the mating flight, the spermatheca (sperm bladders) are filled with up to one million sperm cells. In a normally developing population year by year, this supply is enough for about four years. However, few queens live that long.

In an intact colony, an intact queen 'pollinates' the cells on a comb side with virtually no holes. A breeding nest without holes is therefore an unmistakable sign that the queen is fine and the colony is healthy. Either the queen or a disease of the brood may be responsible for a brood nest with persistent holes.

It may also happen that the bee bread stored mainly at the edges of the (still growing) brood nest prevents the development of contiguous breeding surfaces. This is usually only the case in the long term and has no influence on population development.

It all comes down to the diet

It takes about three days for the larvae to emerge from the eggs. The larvae are immediately fed by the nurse bees. The food is placed next to the larva and is continually replenished so that the larva is never dry. Larger larvae receive more food than smaller larvae.

For the first two days, worker larvae are fed with royal jelly, which the older nurse bees produce in their feeder juice glands. Older worker larvae receive a mixture of feed juice and pollen pulp from the younger nurse bees. This food establishes their further development as a worker. Genetically there is no difference between a worker and a queen. The epigenetic difference only arises due to the different growing up situations. Queens can still be developed from larvae that are one or two days old if they continue to be fed with royal jelly. This fact is useful for targeted queen breeding. We use this fact as well when we make a new colony from just one frame with bees, brood and youngest brood up to two/three days old. In older larvae, such a reversal is no longer possible.

After six days, the worker larvae have grown into thick maggots that completely fill the cells. The brood is then ready to be sealed. The cells are closed by the nurse bees with a porous, air-permeable wax lid. The larva pupates in the sealed cell. The pupal stage lasts 12 days, then the new worker emerges from her cell. When she emerges, she breaks the cell lid herself. Nobody helps her with that.

Many professions needed

The newly hatched young workers have denser and therefore brighter hair than the nurse or fly bees and they are also smaller. With the absorption and digestion of the first food (honey and pollen), their blood circulation starts. The body fills with bee blood and the glands unfold. This causes the body to expand and the bees become larger. The digestion of the food ingested immediately after hatching initiates the preparation for the young bee's work through the activation of the feed glands: feeding the larvae. First, as a young "nurse bee", she takes care of the older larvae, which are fed a mixture of worker jelly and food mash. Later, when her feeding glands are in full production and she is able to produce the high-quality and higher quality royal jelly, she feeds the younger larvae that have just hatched and she also participates in the care of the queen with royal jelly, the "gelée royale", which needs this royal feeding juice especially in spring and summer when this lady lays more than a thousand eggs every day. That is why the queen is either on the combs that still have empty cells to lay her eggs or on the combs where the youngest brood needs to be fed. This you should remember well when you are searching for the queen. The production of feed juice and feed mash requires a lot of life energy. Therefore, the summer bees that work as nurse bees immediately after hatching practice this "profession" for only about one week. They then switch to other activity disciplines that are more determined by the flow, such as comb building and care of supplies and field service, which essentially includes the collection of water, nectar, pollen, honeydew and resin.

All bees present in the colony at that time participate in the necessary regulation of the colony climate (temperature and moisture content in the colony) during the day and at night. All bees pay attention to order and hygiene. Necessary cleaning work is carried out immediately on site by bees. No work is left behind.

In short: every worker bee knows what she has to do at all times. This perfect organisation is based on impressive communication among the colony.

Anyone who wants to understand how the individual worker is able to carry out the many tasks assigned to her must consider her body structure. Knowledge of the development times of worker, queen and drone is important for dealing with bee colonies. This determines, for example, the (annual) schedule for beekeeping during the swarming period, when breeding queens and making splits with bees and brood, and also the schedule for varroa control integrated into beekeeping.

Description of the three honey bee life forms

	Worker bee	**Queen**	**Drone**
egg is placed in a	fertilised worker cell	fertilised queen cell	unfertilised drone cell
larva is fed	initially two days with royal jelly, later with feed mash	continuously with royal jelly	drone feed juice
development time	21 days	16 days	24 days
of which egg	3 days	3 days	3 days
as larva	6 days	5 days	7 days
as pupa	12 days	8 days	14 days
		so: three-five-eight and the queen is made	
when hatching cell seal is	demolished	cut off	cut off
hatching weight	approx. 100 mg	approx. 200 mg	approx. 200 mg
body length	12 - 15 mm	18 - 22 mm	15 - 17 mm
sexually mature after		about 7 days	about 14 days
copulates with	---	up to 30 drones	with 1 queen
lifespan	in summer 2-3 weeks, in winter 2-6 months	3-4 years	1-3 months

number per colony	5,000 to 50,000	usually 1	0 - 1,000
special body parts	honey bladder	maxillary glands	sexual organs
	pollen - collecting device	spermatheca	powerful flight muscles and large eyes
	fat protein body	ovaries	
	feeding juice glands		
	salivary glands		
	wax glands		
	sting and poison bladder	sting and poison bladder	no sting
	scent glands	scent glands	
	poop bladder		

About the origin and life of queens

The queen has less to do than the workers. This assessment only applies if one measures her performance by the number of tasks that workers have to perform and not by the time she spends on her tasks.

A bee colony can raise a new queen for three reasons:

1. If the colony wants to swarm, it creates swarm cells in which new queens become raised.
2. If a colony has lost its queen, for example as a result of an incorrect beekeeping intervention, it only takes a short time (at most 1-2 hours) before it realises that it is queenless. It then breeds new queens from the youngest worker larvae. These queen cells are called emergency cells. A queenless colony can only grow a new queen if it has open, youngest brood. If a brood-free colony is suspected of being queenless, installing a queenlessness test (a comb with open, youngest brood from another colony can check this). If the given brood is further cared for without emergency cells, the colony in question will be completely occupied. Of course, one does not yet know whether the queen of the colony has been fertilised, because then there would be brood present. Therefore, such an occurrence in a breeding-free colony should be checked after a few days. If the queenlessness test is used to create emergency cells, the colony can immediately be provided with a queen. The comb for the queenlessness test is then first removed. The potential of a bee colony to raise a new queen can be used for the targeted breeding of young queens.

3. If a honey bee colony is not (anymore) satisfied with its queen, it can proceed to a "supersedure". This silent queen change is initiated when the queen lays eggs in a pair of queen cups. Unlike swarms, the old queen remains in the colony and is usually tolerated until her daughter starts laying eggs. However, it may happen that the decision as to which queen should rule over the colony does not work out in favour of the daughter, the queen remains in the colony and the colony again proceed to a silent exchange of eggs.

There are no distinguishing characteristics between queens resulting from captive breeding, silent breeding or during swarming, neither in body conformation, nor in body size nor in terms of subsequent productivity strength. A queen is a queen as extensive tests have shown.. Or to put it another way:

A queen is born from a removed queen cell. Next to her a queen cell from which a sister or half-sister has already emerged.

If a queen will be worse or better than others, it does not depend on whether she was born through offspring, silent queen rotation or through swarming. The queens that grew up in swarm cells and exchange cells were destined to become queens from the start, from egg laying. It is different with the emergency cells and with the queens that have grown up in a breeding frame during targeted queen cultivation. The eggs from which they hatched are laid in a worker cell. With normal development they would have become workers. They were only made into queen larvae when this was necessary (for the colony) due to sudden queenlessness. This change of destination, not becoming a worker but a queen, is achieved by feeding exclusively royal jelly. In fact, this is only possible with one-two day old worker larvae. Older larvae can no longer be transformed into queen larvae.

Therefore, a queenlessness test must contain the youngest larvae or eggs and a breeding frame must be equipped with the youngest larvae.

There are also certain differences between the offspring that a queenless colony raises from the youngest brood and those that arise from larvae that are given to a queenless colony during queen breeding.

In the first case, the queenless colony only looks for the youngest larvae from a wide range of candidates, in the second case the colony can only fall back on the few larvae that the beekeeper has given it with his breeding frame.

Feeding royal jelly has consequences.

- The larvae grow larger. The pupae that result from this are also the same. That is why a worker brood cell is too small for a queen larva and especially for a queen pupa. During the care of the larva, the cell is therefore enlarged and extended on the comb surface. Neighbouring cells must make way.
- Queen larvae and pupae develop faster. Compared to that of worker larvae, the larval period has been shortened by one day, and the pupal period has even been shortened by four days.
- The queens live longer. While a worker has a life expectancy of two to three weeks during the summer and two to six months in the winter, a queen can live three to four years, provided she survives the battle with possible rivals, returns from the nuptial flight and, moreover, her colony for three to four years does not give her any reason to initiate the selection (silent queen rotation).

After emerging, the queen needs a few days before she is fit enough for the mating flight. For her nuptial flight the virgin queen flies to the Drone Congregation Area (DCA) with 10,000 to 15,000 drones which is as far away as possible from her nest (at least 2,5 km away) so that the chance that she will mate with a brother is as small as possible. Drones will not fly that far from their hive as they need as much as energy to fly around to wait for a virgin queen to appear. During the mating flight, which only takes place in good weather with summer temperatures, she can copulate with up to

30 drones. The drones are initially attracted by the pheromones of the virgin queen, but as they get closer the drones use their large eyes to follow her movements. The waiting drones line up to mate with the queen and even when she is already mating with another drone. A mating with a drone takes usually less than two seconds. The drones wait without any aggression until it is their turn to mate and die. Because that is what happens with drones that mate. More about this phenomenon will be told when we will talk about drones. The queen receives 70 - 100 million spermatozoa during the mating flight. First all these spermatozoa are stored in her oviducts. However, only 10% of all these 'collected' spermatozoa will be used for the actual storage in the spermatheca. The excess sperm is expelled one or two days after the mating flight. One in five queens does not return to their hive from the mating flight. Upon her return from the mating flight to the colony, she immediately starts laying eggs.

During swarming, during breeding and during silent breeding, the colony always raises several queens, although later not all of them are used, often only one. Growing multiple queens probably reduces the risk. One will come about anyway. Nevertheless, she has to compete with her sisters. If the first to emerge is strong enough, she will be able to kill her rivals in their cradle before they even emerge.

If they missed one in the dark of the beehive or if she has arrived too late, then the decision will follow in a duel, unless a compromise is made that each will rule over part of the colony. Division is generally common among strong colonys. Perhaps that is also the will of the colony. One of those queens will leave the hive with part of the colony and go to search the wide distances. Such a matter will be raised again when discussing swarming.

The egg-laying queen is constantly supplied with royal jelly by the (older) nurse bees. Her care therefore consists of well-pre-digested and high-quality food. In the summer a queen can lay up to 2000 eggs. 2000 eggs weigh about as much as the mother itself. As a result, the amount of food provided to the queen each day if she lays that many eggs must be more than her own body weight. In winter she gets less food. She cannot convert everything she eats into eggs and energy. Nevertheless, she never has to go to the toilet. Not only do the nurse bees constantly feed her, but they also continually preen her and collect her excretions as they arise and carry them further. Woe if the queen suffers from intestinal disease. Then she infects her entire colony.

The cleaning bees not only keep the queen clean, but when cleaning they also absorb the substances secreted by her, which are passed on from bee to bee and thus signal to the colony that she is (still) there. With the death of the queen, the production of these substances also suddenly stops. Due to this sudden loss, the colony recognise the lack of eggs and will ensure follow-up if the latest brood is available. If a broodless colony is affected by the loss of a queen, the colony can only survive with the help of the beekeeper if he comes to the rescue with a replacement youngest brood by means of a queenlessness test.

Vagabonds

The drone's to-do list is considerably shorter than that of the workers. They usually also live a little longer, three to five weeks after emerging. Drones are equipped with clearly stronger flying musculature than the workers; that is why they are larger and are also the better flyers. They have also larger eyes.

Workers are needed all year round, drones only during the summer. Drones are bred again in March, so that even on warm days in April the buzzing of the workers is repeatedly drowned out by the hum of drones. The peak of drone breeding is in May and June and coincides with the swarming season. After that, the workers' desire to rear drones diminishes. In late summer, many colonies no longer tolerate them and drive them away. But that is not always the case. Here and there one can still find a few drones on the combs in November.

Drones are only there to copulate with the virgin queens and they lead a decidedly nomadic life. They only stay in the colony where they grew up for a short time. After orientation flights, they often fly to foreign colonies. They also roam around apiary to apiary. If a colony raises or has raised young queens, then this message is proverbially circulated among the drones in the area. They wait in the colonies on the apiary for the young queens to prepare for the mating flight and they usually immediately leave their quarters for them.

As soon as a young queen fledges, she flies preferably to a drone congegration area about 2,5 km away from her hive where she is chased by a bevy of drones and copulates one after the other with up to thirty drones. The mating is over in minutes. Hence, she can afford to fly further away from her hive than her brothers, because she doesn't have to spend as much time circling in the air than the drones. Honey bees have about half an hour for a flight. Copulation takes place in full flight.

Initially the drones track the virgin queen by her pheromones, but as they come closer they use their large eyes to try to catch her. Once a drone catches the virgin queen, his first task is docking, which he does by flying above her and holding her abdomen (the back of an insect's body) with all six of his legs. The drone then aligns the tip of his abdomen with that of the virgin queen and everts his endophallus (his mating device) by forcing hemolymph (the body fluid of an insect, insect's 'blood') into it. This extreme shifting of hemolymph causes the drone to become paralysed and to begin to fall backwards. His endophallus has several anatomical features that keep it anchored in the virgin queen's sting chamber. The queen aids in this process by contracting muscles in her abdomen, causing full eversion of the endophallus and release of the sperm.

The sperm enters the queen's oviducts and remains there until the queen is finished mating. At this point the still-paralysed drone falls away from the queen and dies. This entire procedure from catching and mounting the virgin queen to the drone falling away takes less than two seconds. Only 10% of all received sperm in the oviducts of the queen will make it to the spermatheca, which holds five to seven million sperm. This means that on average only 3 to 5% of each drone's spermatozoa will be stored into the spermatheca. Even then, it may not ever be used to fertilise eggs.

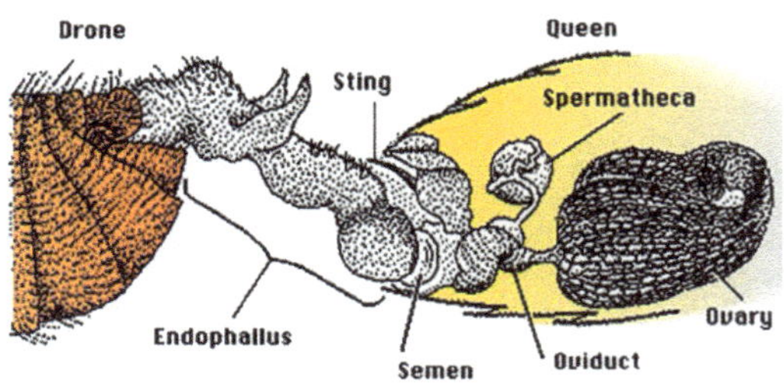

So, any drone that successfully copulates dies afterwards. The others return to the bee colonies for reinforcement and to spend the night to try again a next time. Only one in 10,000 drones manage to pass on his genes to a next generation. Old beekeepers will tell you that nevertheless all drones are important for the harmony in the colony.

It won't work without group work

The summary of the tasks of the three bee divisions in the table only gives an impression of the astonishing performance of the workers. Group work is an absolute requirement for the fulfilment of all these tasks. This applies in particular to 'indoor service' work such as comb building, heating or cooling, storing and preserving supplies and brood care.

Building combs

Honey bees aren't the only insects that can build combs. Social wasps and hornets can do that too. But while they make their combs from paper, honey bees use wax that they sweat out of their wax glands located in their abdomen on the abdominal side.

The wax is made there from the sugars of the honey and is expelled outward in fine, bright plates. The wax plates are handled by the honey bees with their legs and passed on to the mouth tools for kneading. Some plates are lost and fall down. That is why they are easy to find in the waste (on the hive drawer) during the construction season. During kneading, the wax is mixed with saliva and then used as a building material on the colony's construction sites: for building combs and sealing the honey and the brood cells.

Particularly impressive is the regular comb construction. This also occurs if the bees are not given pre-stamped foundations.

If the honey bees are left to build freely, for example in an empty frame, a pure working colony is rarely created. An empty frame hung between two combs is filled exclusively or almost entirely with drone comb by strong colonies during the swarming period. In the late season the colonies also produce worker cells, which are often mixed with drone cells. Only in weak colonies the natural structure is predominantly worker cell structure.

Honey bees only build when there is flow or when they are being fed. Every flow and every feeding should be used to renew the comb house. This is easiest to do if the colonies are given frames with foundations before or during flow or feeding. The pre-stamped foundations ensure that mainly worker cells are built on them.

The tasks of the three honey bee castes

Worker bee	**Queen**	**Drone**
contribute to mutual understanding	maintain the colony's structure	wandering around
feed the brood	kill rivals	copulate with a queen
feed the queen	perform a nuptial flight	
feed the drones	lay eggs	
polishing cells	to swarm	
cleaning other bees		
clear out sick brood		
carrying dead bees out of the colony		

storing and preserving honey		
storing and preserving pollen		
heating		
cooling		
regulate the colony's temperature		
regulate humidity and airflow		
wax sweating		
wax kneading		
build combs		
sealing (capping) brood cells		
sealing (capping) honey cells		
sealing cracks		
monitor the hive opening (entrance and exit).		
repel enemies		
raise the alarm		
chase away drones		
explore flow sources and communicate their location		
record and pass on data from the flow source locations		
collect nectar		
collect pollen		
collect propolis		
get water		
to swarm		

What is put into the combs

A fully developed comb in Zander format can contain up to 3 kg of honey, which is stored in about 2 x 3,200 = 6,400 cells. In easy beekeeping there is no use of different size honey supers. It is wise to choose a hive with about corresponding sizes to be certain at one hand to be able to provide sufficient room for bees and brood during the flow season and at the other not have to handlle too heavy honey frames.

The empty comb from a Zander hive weighs only about 0.1 kg (one ounce) without a frame.

Each cell can be used as a storeroom for pollen or honey or for raising brood. When brood hatches, when the adult honey bee emerges from the cell, she leaves behind the pupal skin that adheres closely to the cell wall. This skin is not cleared away and neither are the excrements of the pupa that are on the cell floor and remain in place between the pupal skin and the cell wall. When the cell is incubated again, the queen lays the egg on or in the remaining skin, the surface of which has previously been thoroughly cleaned by the cleaner bees.

A comb or each cell of a comb can be incubated several times. A pupal skin is added with each breeding round. As a result, the cell volume indeed becomes smaller and the cell walls become thicker. After two to three years of use as a breeding area, the breeding combs are black. Germs can settle between the skin.

Bald brood: a wax moth larva has been at work here. It mines through the cell wall and thus causes the brood to die. The cleaning workers first remove the cell seal (the cell lid, the cap) and then the dead brood. An important feature that may lead to a way to deal with limiting the damage by varroa mites.

Photo Dr. G. Liebig

In addition, droppings and pupal skins are nutritious food for the larvae of wax moths that destroy the comb by piercing it.

A beekeeper must attach great importance to comb hygiene (continuous comb renewal). Anyone who doesn't pay attention can quickly get into trouble. Novice beekeepers are often lucky because they unconsciously do the right thing in the beginning, because they have to do it. They simply have no choice but to keep their colonies on fresh combs, because in the beginning they only have a few or no old combs at all. It takes two to three years until the first combs are old. If they are not consistently removed and replaced year after year with fresh combs, the lack of hygiene can lead to problems and the proverbial beginner's luck is gone.

The development of the bee colony

The honey bee colony is dependent on its environment like no other economically useful animal. This is reflected in many properties, in particular the seasonal development during the honey bee year.

Determining the start of the honey bee year is really a matter of perspective. For the honey bees it can be as the days lengthen and also the weather is nice or when the willows begin to bloom. For the beekeeper it is August when it is time for the honey bees to make winterbees. Then the honey bee year does not start and end like the calendar year with New Year or New Year's Eve, but in late summer when the honey bee colony prepares for hibernation by raising winter bees.

With regard to the development of the honey bee colony, we divide the honey bee year into five phases:

- the spring development
- the swarming time
- the highlight in summer
- the development in late summer and autumn
- the over-wintering

The breeding period in our latitudes lasts approximately from February till November. Weather conditions have a decisive influence. In a mild winter, the honey bees can continue breeding without interruption. However, the breeding nest remains significantly smaller than in the summer.

Spring and spring development

Depending on the location and the weather, the honey bee colony starts breeding between mid-February and mid-March. In mild weather the brood nest expands quickly, so that usually in March the number of brood cells exceeds the number of bees. With the hatching of this brood quantity, colonies begin to grow. The brood cell to bee ratio can already be greater than 2:1 by mid-March. This can always be the case sooner in climate-dependently warmer places than in colder places.

Whether there is little or much breeding depends on the weather and population strength. If winter lasts until March, the colony will wait and see. After a long winter, spring often starts at record speed, causing vegetation to develop explosively. The honey bee colonies then follow the development of the crops.

If winter ends early and spring-like temperatures prevail as early as February, this will affect both the vegetation and the honey bee colonies. If the hazel blooms earlier than normal, the honey bee colonies will also start breeding earlier and their breeding nests will expand more quickly.

Unlike vegetation, the honey bees do not let themselves be stopped by later cold periods. Only if they are of a very long duration do the honey bee colonies shrink the brood nest, usually due to a lack of pollen. In normal cases they continue to breed even when it gets cold again. They also continue to grow in the spring when cold weather halts vegetation development. It results in the colonies being very strong and vigorous when the fruits and the meadows begin to blossom. In or after a spring with changeable weather in March/April, the honey yield from the spring flowering crop is often higher than average, but only if the consumption of the spring flowering crop in May is not hindered or even prevented by persistent rainy and cool weather.

The colonies on an apiary usually develop very differently in the spring. The rule applies: the stronger the colonies were during wintering in, the stronger they are during wintering in, the more intensively they start breeding, the more vigorously they grow, the more nectar and pollen they collect, the sooner they become in swarm mood.

The size of the brood is highly dependent on the number of honey bees. That also makes sense. A colony of 10,000 honey bees strong can provide more brood than a colony of 5,000 honey bees and less than a colony of 20,000 honey bees. Naturally, weaker colonies produce more brood than stronger colonies, which can lead to the weaker colonies catching up in May when the big flowering begins, and can look almost as good as the colonies that have emerged better from the winter. They can do that on their own. Any help from the beekeeper is not necessary. This ability turns young colonies, who are generally weaker in and out of season than older colonies, into "racehorses".

This colony overwintered in one breeding box and at the beginning of April was given a honey box as a second box above a queen excluder. The result: the bottom box is filled with drone cells. Just one breeding box with ten (Zander) frames as breeding space is not enough.

Photo Dr. G. Liebig

The beekeeper does not have to do anything or little for the development of his colony. However, he must ensure that the colonies do not get hungry in the spring and that their enclosure is expanded in time so that this does not become too small.

Feed consumption is extremely high in the spring and especially when the weather is changeable during March and April. The stronger colonies or the stronger breeding colonies consume the food the most. A colony can become more than 10 kg lighter in March/April before the flow begins. Obviously, food consumption to this extent is only possible if so much food is available to the colony.

Growing and wanting to swarm

Growing in the spring is one of the innate characteristics of the honey bee colony. The bee colony's urge to grow in spring is natural and (more than) vital and therefore particularly pronounced. As many workers as possible are needed so that nature's temporary supply (flower production or flow) can be optimally used. Only those who collect enough supplies can survive the winter. The honey bee colony is well equipped to achieve this task:

- through the diligence of every bee and
- due to the number of workers that only grows after wintering for economic reasons.

Growth is not only necessary because of the acquisition of supplies. Growth is also the possibility of multiplication whereby as many offspring (swarms) and drones as possible will arise from a colony.

A strong honey bee colony wintered with sufficient supplies will develop vigorously in the spring, produce a lot of nectar and pollen and will want to swarm.

Depending on population strength, location and weather conditions, the queen's laying capacity reaches its peak between mid-April and the end of May. Every day she can lay 2,000 eggs and sometimes even more. If she does this without interruption over a period of at least three weeks, then 42,000 cells will have been incubated. More is not possible.

The eggs first produce larvae that need to be cared for and then workers that want to take care. The larval life is set at six days. The total service duration time of the workers is flexible, it can be 10 days, but also shorter. At the same time, it must be remembered that one worker can care for several larvae.

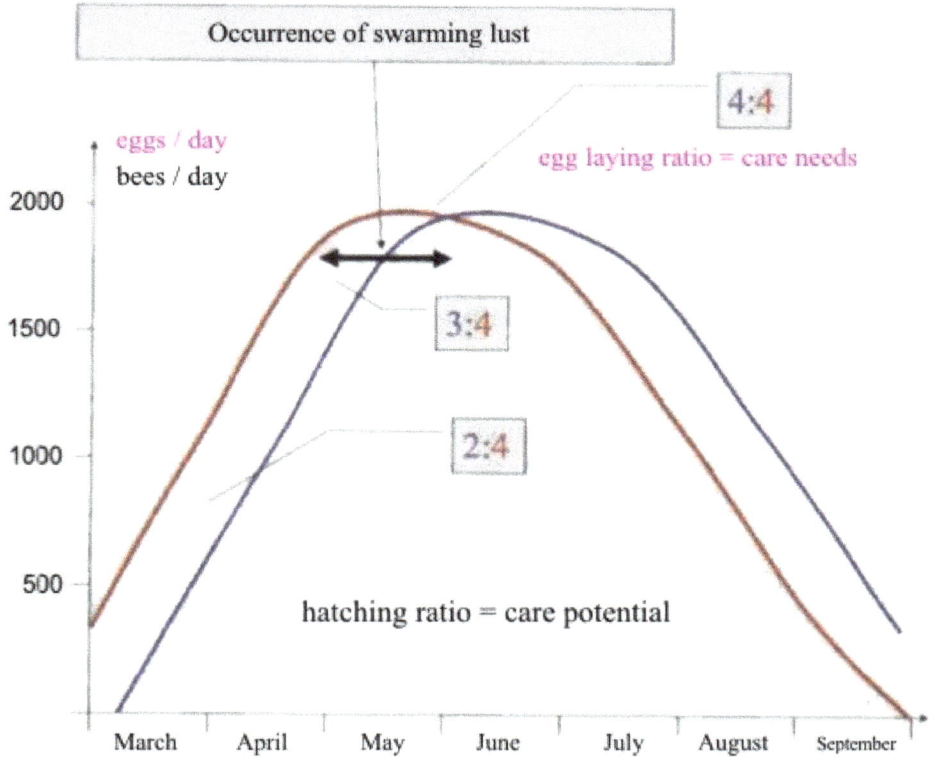

Swarming is generally limited to the period May/June. It is part of the nature of a honey bee colony. The cause of swarming can be deduced from population development. By forming splits with bees and brood, the occurrence of swarming behaviour is prevented by removing "care potential" before the colony becomes unemployed and starts to have "swarming thoughts".

Picture after Dr. G. Liebig

As long as the queen lays more eggs day by day, the young workers are prepared to do care work, this is no longer the case once the queen has reached her maximum laying capacity. The "demand for care" then remains constant, albeit at a high level, while the "supply of care" continues to increase. Three weeks later, about 2,000 honey bees emerge every day and a few days later they want to take on their care tasks as young workers. If a lot of "care supply" meets relatively little "care demand", the "swarm lust" can awaken. That can be expected first in colonies that have emerged strongly from the winter. Moreover, especially during good flow, the queen's laying capacity may be limited by a brood nest filled with honey. As a result, the "care demand" decreases while the "care supply" continues to grow with the emergence of the brood in the first large brood nest. An increasing number of workers can no longer get rid of their feeding juice. If the beekeeper does not intervene, the bees will help themselves and the meaningfully co-ordinated steps will lead to swarming.

3 kg swarm = 2 kg bees + 1 kg honey

First, more and more play cells are made, preferably at the edges of the breeding frames. As soon as the first play cell is filled with an egg, is the colony in swarming mood. This has consequences. The gathering effort decreases (slightly) and all construction comes to a complete standstill, because the swarm bees prepare for the journey into limbo. The queen is in the hive chased by the workers so that she gets in flying condition. Also the older bees stop their work and pack their bags, which means they build up a protein-fat storage body in their abdomen and fill the honey bladder with about 50 mg of honey. With the help of these supplies, comb construction can immediately start in the new housing and the first brood can be cared for. Immediately after sealing the first swarm cell, the swarm is ready to start. The swarm leaves if or as soon as the weather is nice. The swarm is composed of older workers and flying bees and also attracts flying bees from other colonies. The exodus takes about ten minutes. Every minute, more than 1,000 honey bees leave the hive, including the old queen. First the bees buzz around in the air until they gather nearby in a suitable spot, usually on a branch of a tree, and form a swarm cluster. Immediately after the exodus (and sometimes even earlier), scouts fly out to look for suitable housing. Sometimes a swarm spends the night hanging on the branch and does not move on until the next day. This ensures that all scouts come along and that no one is left behind. The beekeeper has the opportunity to catch the swarm beforehand in the evening or early in the morning. However, it is more common for the swarm to move on already on the day of the exodus. Then it is usually lost. Anyone who arrives on site in time for the exodus can try to catch the queen when she appears in front of the flight hole or before she takes to the air from the fly hole. One can also place a (piece of) queen excluder in front of the fly hole so that the queen cannot join the swarm and is then forced to return to try again soon or else a little later. It goes without saying that the "queen excluder for the fly hole" is completely inappropriate for swarm prevention. In this way, the drones are also trapped, which in their urge to go outside become trapped in the queen excluder and thus clog the entrance. It is better if there is no exodus of a swarm at all.

The queen can also be clipped on one wing. This has no effect on her laying capacity, she simply becomes unable to fly. She cannot follow the swarm during its exodus and usually ends up "in the grass" in front of the fly hole without any substantial accompaniment. When the swarm bees notice that the queen is not with them, they return to the hive and wait there until the first queen(s) emerge and then set out again. Because up to that time more than a thousand bees have completed the pupal stage since the failed exodus, such a "singing swarm" is much stronger than the previous front or main swarm.

Anyone who clips the wings of their queens only gains a few days of time. If the beekeeper does not use that time to break the swarm cells, he will lose more.

How does one catch a swarm?

Most swarms settle on a tree branch, rarely at eye level. A ladder is usually required. Standing on the ladder, the swarm catch basket is held under the swarm and the swarm is knocked into the collection container or the catch basket (bucket) with a striking blow with a sweeper or by vigorously shaking the branch or by tapping on the branch. The aim should be that at least the queen falls in and stays there. Whether this was successful can only be determined after descending the ladder.

The swarm collection container (basket, bucket, skep) is placed aside in the shade and covered with a queen excluder and the lid of the collection container without closing it.

A swarm with a queen remains in the collection box or catch basket, forms a cluster there under the queen excluder and attracts the honey bees that have not yet been captured, which gather on the queen excluder and mark the swarm's new home with intensive fanning. The swarm can then be placed on the spot in a beehive equipped with foundations (and a honey frame) and a few days later treated with oxalic acid against varroa mites.

Dr. G. Liebig catching a swarm in an apple tree. The swarm catcher sweeps the swarm into a large bucket with a sweeper while standing on a ladder.

The captured swarm with the queen has been knocked into a honey bucket. The queen excluder placed on top prevents it from flying away.

Photo's Dr. G. Liebig

A swarm without a queen leaves immediately and returns to the branch with the uncaptured bees and the queen. However, the captured drones under the grid cannot follow the swarm. When the swarm cluster is hanging on the branch again, you can try again.

The most accidents in beekeeping occur when catching swarms, because standing on the ladder requires three hands. Therefore, in "high" cases it would be better to refrain from catching a swarm.

Shared risk

The descendants of the swarmed colony bear a distinct risk.

The swarm that has moved away must find a home and build up honey and pollen supplies as quickly as possible to survive bad times, especially winter. If the weather and flow work out well, they will succeed. The conditions for this are available from the swarm. The swarm has mainly recruited forager bees (fly bees) and older hive bees (house bees) that already have experience with brood care and collecting nectar and pollen. The queen of the swarm has already been fertilised and can immediately start laying eggs as soon as the first combs have been built in the new housing. Three weeks later the first brood emerges and the colony begins to grow.

The swarmed colony can release one or more subsequent swarms after the departure of the preliminary swarm. That depends on the behaviour of the young queens. The first queen to emerge usually tries to kill her still-sealed sisters. She bites open the side of the swarm cells and then stabs the pupae inside to death. If she succeeds in killing all rivals, she will be the sole ruler of the swarmed colony, which has been strengthened again since the departure of the swarm until the emergence of the first young honey bees.

Depending on the egg-laying capacity of the swarmed old queen, up to 2,000 honey bees still emerge every day during this period. Maximum time between the departure of the pre-swarm, which is established at the earliest as soon as the first swarm cell is sealed, and the emergence of the queen that has grown up there, lies for eight days and in that case more than 15,000 honybees can emerge. The swarmed colony loses few bees during this period because the collection of bees is difficult due to the missing flying bees stocks have almost completely ceased. It can then easily happen that the honey bees of the reinforced residual colony prevent the first emerged queen(s) from killing her sisters and prefer to walk out with her or with them as an after swarm or after swarms. After-swarms are significantly weaker than the main or pre-swarm. They carry a double or even triple risk: they have no housing, the queen is still unfertilised and - because of their lower strength - they have to make significantly more effort to survive until autumn to grow into a population that can hibernate successfully.

The question why the old queen leaves and not her daughter.

Contrary to other eusocial insects (like ants, termites and stingless bees) the old queen of honey bees leaves the nest and not the or a young queen, her daughter. In other eusocial insects, a term introduced in 1966 by Suzanne Batra to describe, for that mater, the nesting behaviour of another bee species than the honey bee, the old queen stays in the nest where she has started her reproductive life. In honey bees the old queen takes the risk, so to say, to start a new colony in uncertainty. It is probably more of a honey bee colony drive to leave the nest with the ruling queen than that it is her wish. So, apparently a matter of developed genes. As it is known that in the early evolutionary stages of the development of the honey bee this leaving the nest by the old queen was not the case, the search is than for the evolutionary moment that this habit was established, what did trigger this phenomenon, why has it been established? Most certainty this phenomenon must be benificiary to the honey bees and and must have been once it occurred. The advantage of this phenomenon must lay in the chances of success both for the remaining colony with the new queen and the new colony with the old queen. This results from the perspective of the 'survival of the fittest' or 'what is realised and could reproduce from the possibilities of life that were available at a given moment in time'. Data about this are scarce. If looked at the succesrate of overwintering of honey bees in the wild as has been done and reported in 1987 and 2017 by Tom D. Seeley the figures are not encouraging for the swarms (about 80% on average) and most certainly when these are compared with those for the overwintering 'old' nests (about 25% on average) in North-America. These data were gathered in an area in the world where the honey bees did not suffer from varroa infestation, for that matter. Swarms in areas were varroa is present, as is nowadays the case, do most of the time not survive the winter. Also a very good reason not to let your colonies swarm.

When the swarm begins the new colony they must start from scratch so-to-say: they have to build combs and they must gather sufficient food to overcome wintertime. The only 'advantage' would seem that the old queen can lay eggs almost right from the start while in the 'old' nest it will last almost two weeks before the new queen is ready to produce eggs if she returns from her mating flight (80% chance). A difference of three and five weeks respectively on average for new bees to emerge, and that in a time that the seasonal flows are decreasing. Nevertheless the swarms do have poor results in surviving wintertime as the figures show. So, when the mutation with the preference that the mother queen should leave her nest in the reproduction phase occurred her daughters started to leave the nest. If this mutation was dominant also daughters from her sons (the drones that emerged from her unfertilised -haploïd- eggs) with that dominant mutation would immediately have shown this habit of leaving. So this preference is only benefitting the younger queen by having better chances this way to survive winter time if she stays in the nest where she was born. If, as a rule, in the evolutionary progress of life the emphasis is always on establishing a next generation and as successful as possible as well, this preference is than understandable if the mutation is spread within the honey bee species as a whole in a relatively short time which must have been the case referring to the low survival rates of swarms, anyway and also compared to those of the nests left behind and referring somwhat perhaps also to the high recombination rate of honey bees. Still, all this is hypothetical.

Another hypothesis has been brought forward by Jacques van Alphen (2024) who is pointing out that the queen leaving the nest is part of the defence strategies of honey bees against germs (pathogens). During the seasonal development of a honey bee colony there is no genetic change compared to the genetic changes of the pathogens present whereby a highly virulent pathogen strain can arise against which the honey bees have no defence at that time. If the old queen remains in the nest during and after the reproduction period the following year the bees of that colony will increasingly suffer from diseases. With the return of a young queen from her mating flight to the nest, a complete new set of genes will come into action causing phenotypes that are new to the pathogens and thus the honey bees have better opportunities of surival. This could be as well one of the reasons that swarms with the old queen and the already adapted pathogens to her bees present and to come have such a low survival rate already without the infestation by varroa. Tests should be made to get some answers and proof in these matters. Also some light should be shed than on what strategies other eusocial insects, especially those with lasting wings, have in their defence against pathogens.

For beekeepers this all is one more reason to replace the queen with a younger queen every year.

Development in summer and autumn

A colony that has not swarmed reaches its peak in the summer. Particularly strong colonies can number around 50,000 honey bees in June and have a brood nest of more than 40,000 cells, but those are exceptions. Most colonies are weaker. The normal colony strength in summer is about 30,000 honey bees. Their breeding nest may have once contained 40,000 cells recently, in May or June, but has been shrinking since then. Until the end of July, the size of the brood decreases only slowly. Breeding activity is drastically curtailed in August and September. As early as mid-August, most colonies usually have fewer than 10,000 brood cells. Moreover, the stronger colonies shrink their breeding nest faster than weaker colonies, which tend to continue breeding in late autumn and winter than the stronger colonies, many of which are already brood-free in October. Apparently the weaker colonies have an external reason for this. Their queens do not stop laying eggs until or during the first frost. Three weeks later these colonies are breeding-free.The majority of honey bees that emerge in the fall develop into winter bees. The steadily shrinking breeding nest ensures that most of them are no longer burdened with labour-intensive and resource-consuming brood care. That guarantees a longer life. The winter bees prepare for winter and the coming spring. They produce a fat-protein cushion in their abdomen, just as honey bees do in preparation for swarming. Important proteins are stored so that the end of winter brood care or the production of high-quality feed juice is possible, even if no pollen or bee bread is immediately available.

And they practice winter clustering. On the cold autumn nights they crowd together in a winter cluster that they warm in autumn or winter, but leave on sunny days.

Such flying days are mainly used to empty the poop bladder. Security for everything. Who knows how long the next frost period will last.

The colonies can survive a flight and toilet break of about 3 to 4 months undamaged, if during that time they:

- be put on a low-sugar diet
- do not care for brood
- only burn honey for heating

In this respect, sugar food honey is considered more beneficial for bee health than real honey; because real honey clearly contains more minerals that remain behind like slag when dressing and burden the fecal bladder.

There is a big difference in quality with real honey in this regard. Flower honey has a significantly lower mineral content than honeydew honey. Honeydew honeys are therefore completely unsuitable for overwintering.

Characteristics and phases of population development in brief

With the figures and the tables below:
The development of a honey bee colony that does not swarm after its start in May as a young colony (that has emerged from a split with bees and brood or from a fertilised young queen colony) until the spring of the following year, depicted schematically. The variables and phases of population development are indicated with numbers and letters. The numbers refer to the bee curve, the letters the breeding curve. A brief description of these variables and phases as well as the most important rules of population development can be found in the following tables.

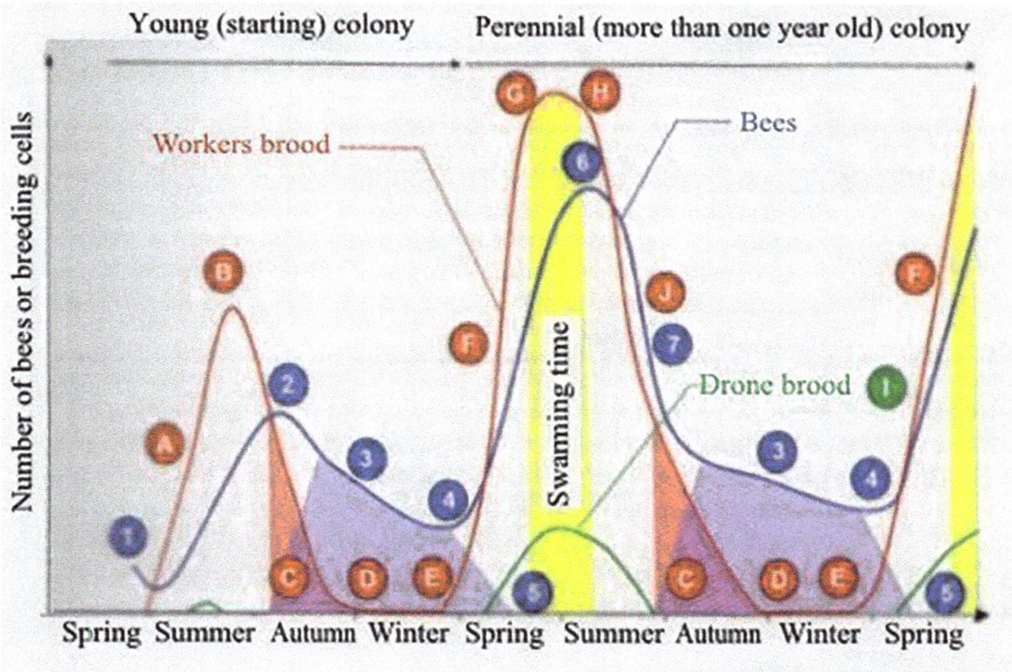

Picture after Dr. G. Liebig

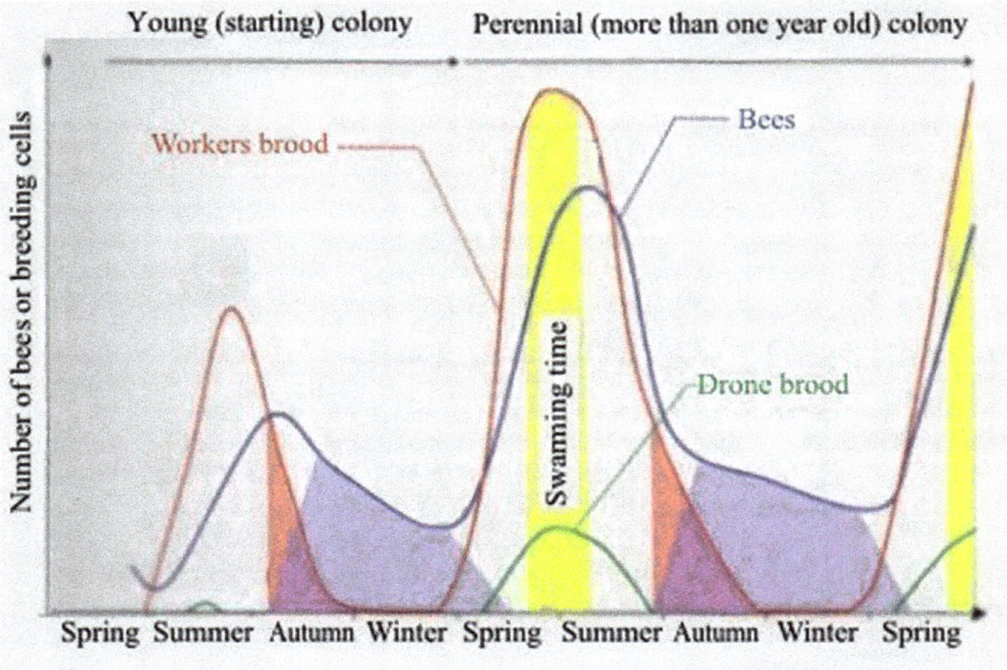

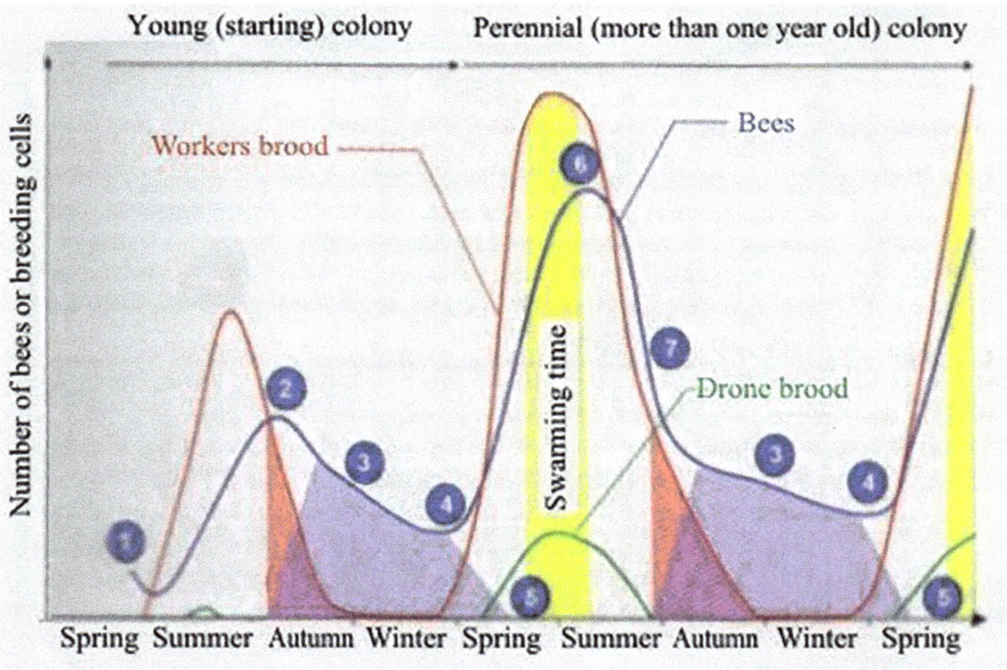

The most important rules of the development of colony strength of fertilisation colonies up to the second wintering in phases 1-7

Phases and variables		Rules and relationships in brief
1	Formation strength of the young queen fertilisation colony	The stronger the fertilisation colony, the more certain is the fertilisation success of its queen. This is generally around 80% from May to August, if the fertilising colonies are 1,000/2,000 honey bees strong, and more than 90%, if the fertilising colonies number more than 5,000 honey bees and in the period of 1-3 weeks after hatching of the queen and there will be nice weather for a few days.
2	Maximum number of honey bees of the young colony	The maximum number of honey bees of the young colonies resulting from the fertilisation colonies usually occurs in September 3-4 weeks after the breeding maximum and is usually significantly below that. The life expectancy of the summer bees raised in the young colonies is clearly less than three weeks. The bee maximum is not or only weakly correlated with the formation strength of the fertilisation colonies.
3	Wintering strength of a young colony or perennial colony	The wintering strength is the number of honey bees of a colony in October. This is therefore assessed 1-2 months before the actual start of winter. Young and adult colonies should have at least 5,000 honey bees by the end of October, regardless of the status of the brood, so that they will certainly survive the winter. With a wintering strength of >5,000 honey bees (and if the other guarantees for a certain overwintering are met), the mortality risk is 3%, with a wintering strength of >7,500 honey bees only 1%. Weaker colonies (<5,000 honey bees) should be reorganised or united. Wintering strength is not or only very weakly correlated with brood production in autumn.

4	Overwintering and outwintering strength	Honey bee colonies survive every winter if they enter the winter strong enough (>5,000 bees), healthy (varroa deposition rate of the winter bees <10%) and with adequate food supply. The wintering strength is very closely correlated with the wintering strength. On average, over several years, a honey bee colony loses approximately 30% of its strength in winter (between the end of October and the beginning of March). Honey bee colonies can also grow in winter, but of course only if January and February are very mild. They than therefore consume more food.
5	Life expectancy and end of winter bees	Using a late nectar or pollen collection of Phacelia, mustard or ivy flowers does not cost winter bees and does not result in increased breeding activity. Sprinkling the tightly clustered winter bunch once with oxalic acid (around Christmas) does not cause any harm if the colonies are opened and treated during severe frost*. The "cold wintering" above an open mesh bottom box ensures a "quiet winter seat". This positive effect is further enhanced by an empty container placed on top. It has been proven by extensive and various tests that there is no need for fear of thermostress or any other myths regarding such. Honey bees keep themselves warm, not the hive where they are housed in. Ventilation is very important.The generational change of winter bees and summer bees usually takes place in March with the hatching of the first brood. * Sprinkling oxalic acid can be skipped if the colonies have been treated with oxalic acid evaporation in the autumn that does not harm to honey bees at all.

6	Bee maximum of the production colony	Well-cared for production colonies grow between 30,000 and 40,000 bees in the summer. It is very rare for a colony to have more than 40,000 bees. The bee maximum usually sets three weeks after reaching the breeding maximum or a little later and is usually significantly below that. The life expectancy of the summer bees that emerged during the growth phase is clearly less than three weeks.
7	Colony strength during late summer and autumn care	In August and September, the perennial colonies clearly lose strength, even if the honey bee number does not decline as drastically as the brood size previously did. Adequate protection of the winter bee brood against excessive varroa contamination is the extremely strict condition for sufficient wintering strength and a long lifespan of the winter bees.

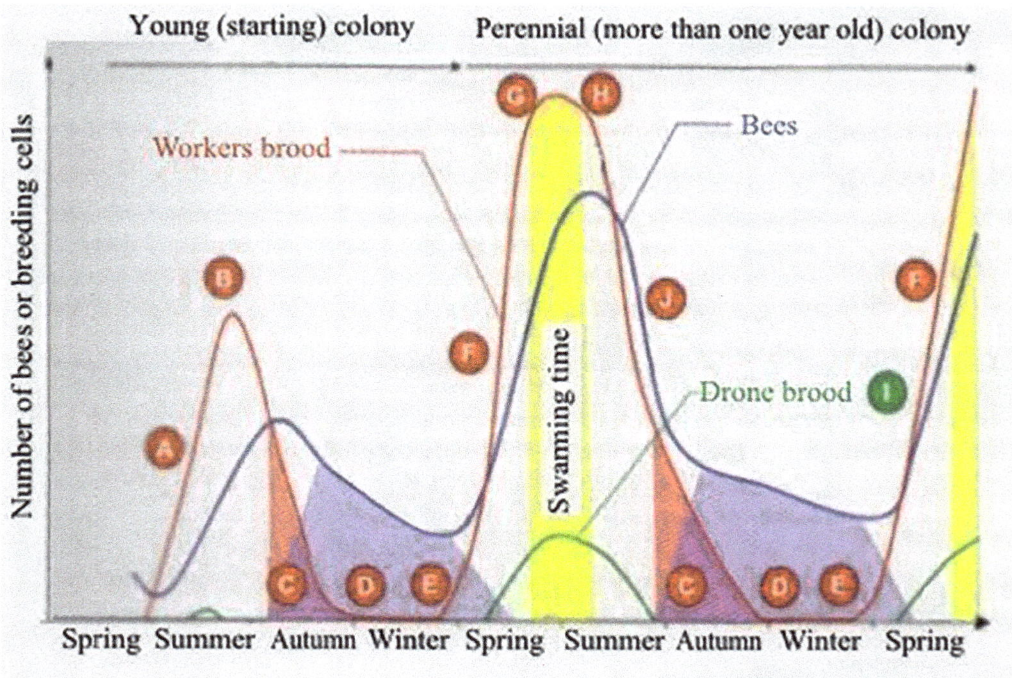

The most important rules of the breeding behaviour of fertilising colonies up to the second wintering in phases A-J

Phases and variables		Rules and relationships in brief
A	Starting brood production of the young colony	Immediately after returning from her mating flight, the queen starts laying eggs. Her laying production is initially determined by the strength of the fertilising colony. The following applies: the more bees, the more brood. In addition, the weaker colonies breed relatively more brood. The brood cell to bee ratio can be up to 10:1 and also depends on the weather. In case of cold waves ("Ice Saints", "Sheep Cold") or lack of pollen, brood cannibalism can occur. First, the youngest brood is eaten. In very weak colonies, the daily laying production of the queen before the hatching of the first brood may decrease.
B	Breeding maximum of the young colony	The earlier in the season and the stronger the fertilising colonies are formed, the earlier and higher the breeding maximum occurs. The breeding maximum of young colonies resulting from fertilisation is generally reached in the second half of August and is then between 15,000 and 25,000 cells. The varroa infestation of the young colonies in early autumn is generally correlated with their brood production achieved from May to August.
C	Growing up of winter bees in young and adult colonies	Winter bees emerge in the autumn or are incubated in the autumn. The first ("winter bees emerge in the autumn") applies more to the stronger or earlier emerging colonies. The second ("winter bees are incubated in autumn") concerns the weaker and longer-breeding young colonies, but also perennial colonies that have suffered from high varroa infestation in late summer. The appearance of winter bees is associated with the decline and cessation of honey bee colony breeding activity in late summer or autumn. Winter bees apparently live longer because they do not nurse brood immediately after emerging.

D	Breeding in winter	The breeding behaviour of the honey bee colony in winter is determined by the day length and the weather conditions. As a rule, the colonies leave the brood in late autumn, the stronger colonies (>10,000 bees) usually without external cause. The weaker colonies only stop breeding completely during a cold period associated with night frost. Three weeks later, all colonies are generally breeding-free.
E	Start of breeding towards the end of winter	Around the turn of the year, the willingness to breed increases again. The breeding size in January and February is very dependent on the weather. A change between warmer and colder periods also affects the size of the breeding nest with the respective delay waves.
F	Breeding progress in spring	As soon as the weather conditions allow a more or less continuous influx of pollen, the bee colonies start to enlarge their breeding nest. The number of brood cells generally increases linearly from March to April/May. As early as March, the brood cell/bee ratio reaches or exceeds the value of 1. Approximately three weeks later, significant bee growth begins. Honey bee number and brood size mutually condition each other during the growth phase. The following applies: "the more bees, the more brood" and "the more brood, the more bees". This coherence is (only) lost when the breeding maximum or maximum laying production of the queen has been reached. In addition to a lack of pollen, cold waves in the spring can cause stagnation or limitation of breeding activity.
G	Breeding maximum of the production population	Depending on wintering strength and weather conditions, the breeding maximum can occur in the spring between the end of April and mid-June. It rarely exceeds 40,000 brood cells, which corresponds to a laying production of the queen of on average 2,000 eggs per day. Continued nectar supply in May/June leads to a generally temporary breeding restriction due to the honeying of the breeding nest.
H	Brood production in the summer	A persistent nectar supply in May/June leads to a generally temporary breeding restriction due to the honeying of the brood nest. Every "breeding kink" is followed by a "bee kink".

I	Growing up of drones	Drone brood is grown from March to July, with a few cells here and there in late summer and autumn. The drone brood curve follows the worker brood curve. The maximum of that curve in early summer is about 5,000 cells. The relative varroa contamination of drone brood is on average approximately 8 times higher than worker brood. Cutting out 3-5 drone brood frames from late April to late June reduces varroa infestation by about half.
J	Shrinkage of the breeding nest in late summer and autumn	In perennial colonies, the breeding nest start to shrink at the end of July. Until mid-September, the queen's daily laying production drops steadily from 2,000 to 200-300 eggs. With a delay of about 2-3 weeks, the bee number also steadily declines.

Late summer and autumn

When considering the condition of honey bee colonies in late summer and autumn, a distinction must be made between young colonies and overage colonies. Young colonies arise from layers or fertilisation colonies that were formed in April/May. These are generally weaker than the older colonies (the overage colonies). However, they can continue to grow until September, while the older colonies clearly lose strength (in terms of number of bees) until autumn.A normal strong colony will have about 20,000 bees and a brood nest with about the same number of brood cells in mid-August, before the beekeeper begins late summer care. After shrinking, it usually sits in two brood boxes that the colony fills completely. If the beekeeper checks the strength of such a colony with a look from above (through the foil) and a second look from below (e.g. through the entrance), but best with a tilt check, then in both the upper brood box and the lower one all the spaces (seams of bees) between the frames - also those between the combs and the box walls - are occupied. In the lower box or in the bottom box (in which the entrance is located) the bee colony hangs over 4 to 6 frames after the end of the (daily) flight period. The brood nest shrinks continuously by 0 - 5,000 cells until mid-October. At the same time, the number of bees also decreases. Usually, the number drops to around 10,000 bees. Some colonies retain their August strength, but there are also those that, despite high brooding performance, are clearly weaker than 10,000 bees. This development is the rule. The strength limit for a successful wintering is around 5,000 bees.

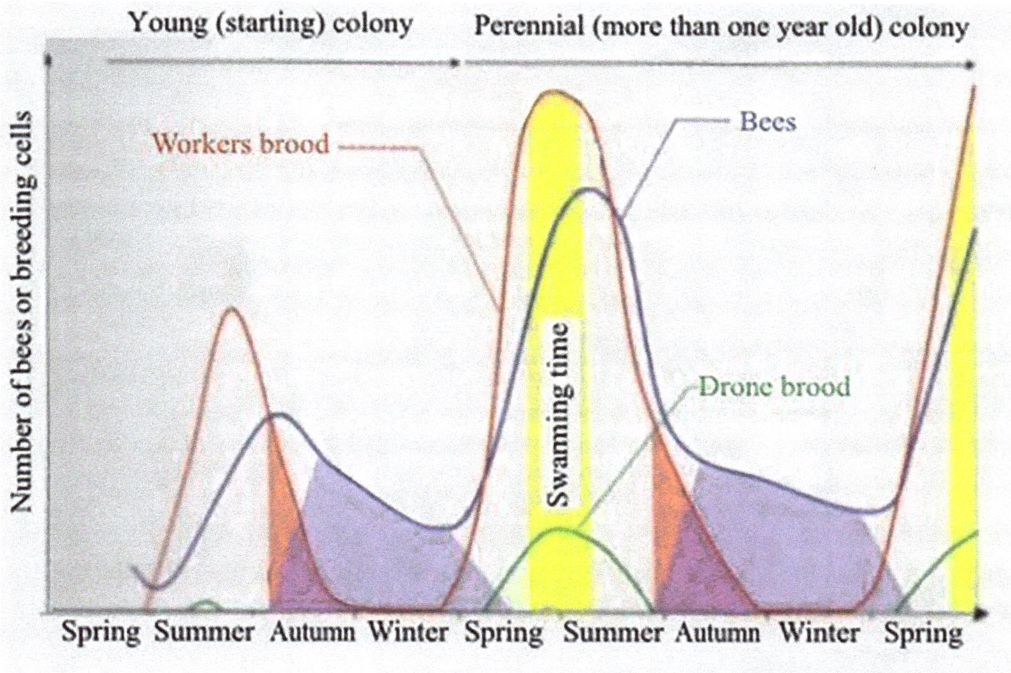

Picture after Dr. G. Liebig

Winter and hibernation

During the winter, the colony forms a winter cluster. The honey bees huddle together in cold temperatures. A colony of normal strength then occupies four to six seams of bees, but not over the entire length. If the (stronger) honey bee colony has filled two brood boxes at the beginning of August, then it needs less than a quarter of the space that is available to it in the winter. In the unoccupied frame seams it is almost as cold as outside in front of the flight slit. Don't worry, the colony does not freeze even then. It can only freeze when it has no more food. As long as the colony still has sufficient supplies, the honey bees provide warmth by burning the sugar from the stored winter food and converting it into heat.In the winter cluster, the temperature is around 20°C as long as the colony has no brood. Once breeding starts, more heating is required. The brood needs a temperature of approximately 35°C for healthy development. Of course, only the small brood nest in the core of the winter cluster is kept warm. The honey bees that sit on the brood-free combs are cooler. The honey bee fur of the winter cluster has a good insulating effect. It protects against unnecessary heat loss. The space that is not occupied is not heated. This saves food and it would also be absolutely unnecessary. The winter food stored there is preserved cold during processing in late summer, stored with a low moisture content and well sealed with cell lid wax. It cannot spoil, even if the feeding surfaces are left by the honey bees. On the contrary: stored cool, it stays fresh longer. The unoccupied feed combs must not be removed by the beekeeper and replaced by 'heat' constructions. Anyone who wants to keep their honey bees warm in this way achieves exactly the opposite. Honey bees also need food in a warm hive. If keeping it warm entices it to start breeding earlier, then it needs even more food than in a cold, apparently poorly insulated hive. The honey bee colony can only consume as much food as it has available. The amount of food that is stored on the closely packed winter cluster combs is insufficient for the duration until the start of the nectar flow in the spring. The colony must be able to rely on the reserves in the combs that are not occupied by the winter cluster. Otherwise, it will run out of fuel in the winter or at least in the spring.In winter, a honey bee colony is careful not to lose contact with its food supply. To this end, it moves its position in the course of the winter in such a way that the cluster always remains in contact with the cells filled with food that are usually located above and/or behind the seat of the honey bee cluster after feeding (sugar a while after the honey harvest) seen from the entrance, or behind the seat of the honey bee cluster. The honey bee cluster is divided into slices so-to-say by the combs. The bees in the upper area of these slices sit on or next to the cells filled with food that they take from there and pass on further down their seams without having to leave the cluster. In the lower, larger part of the cluster, the comb cells of the comb surfaces occupied there by honey bees sitting there closely together are usually empty. In winter temperatures the honey bees crawl head first into the cells, causing the cluster to sit very tightly together. In a comb alley of ten by ten cm squared, About a thousand honey bees live together in a seam measuring 10 by 10 cm square. If a bee cluster in a brood box occupies at least four comb seams over the full length and about half the width of the combs during frost, such a colony counts more than 5,000 honey bees. During the winter, the winter cluster moves upwards and/or backwards in the comb seams it occupies. If these combs are empty, the cluster must either get food from previously unoccupied combs or switch comb seams. If the colony fails to do this, it will lead to a "food demolition". As a result, a colony will starve to death next to combs filled with food. Weak colonies are more likely to suffer this fate than stronger colonies, and especially when breeding has started at the end of the winter and the honey bees are "banished" from the brood nest that has been constructed. Therefore, during the late summer and autumn care in which the colonies are prepared for the winter, care must be taken to ensure that the colonies are strong enough and that they enter the winter with a sufficient supply of food. During the winter, the honey bee number decreases. The dying bees try to leave the colony. They also succeed when it is not too cold. Only at very low temperatures (well below 0°C) do they not get far and fall into a heap in the lower box as fatalities. In normal weather conditions with temperatures around 0°C or higher, the honey bees approaching death fly away and do not return. The beekeeper does not notice that his colonies are getting weaker. Typically, the strength of a colony decreases by a quarter or a third in the winter. For example, out of 10,000 bees, about 7,000 bees can survive from mid-October to mid-March before the first brood hatches in larger numbers. That would be normal. The deviation from this norm is determined by the location (climate), by the weather, by the beekeeper and by diseases. Sick or improperly treated colonies can die during the winter. By far the most important disease that is consistently responsible for colony losses in the winter is varroa. Sometimes it also happens that colonies become stronger during the period mentioned. This is particularly possible in mild winters on climatologically favourable locations, when the colonies start breeding in February or even January and this early breeding results in early growth. Breeding and the associated increased heating not only result in increased food consumption. The stronger metabolism, in particular the conversion of pollen into food juice, also leaves more waste products behind that fill the workers' excrement bladders more quickly. Their storage capacity is not unlimited. The honey bees burdened in this way must have the opportunity for a cleansing flight in shorter periods of time (because they do not want to contaminate the hive) in order to be able to relieve themselves outside the colony. If the weather conditions do not permit this, it may happen that some honey bees, usually the sick bees, defecate in the colony. The consequences for such colonies can be catastrophic. Often this excrement contains germs (pathogens). When cleaning bees then remove these excrements, which usually can only happen weeks later, they are infected and also become ill and can then transmit the disease to their sisters and half-sisters. The disease can then spread epidemically in the colony. Sick bees die considerably earlier. The colony first stagnates in its development, then becomes visibly weaker and can even perish, even while it is fights against it. Self-healing is possible, but only if bee-friendly environmental conditions prevail. The beekeeper cannot help a sick colony, or can only help it to a limited extent. Mild weather in the middle of winter can also have a disastrous effect on overwintering, especially if it lasts for weeks and is followed by a period of frost; because the mild weather can trigger the start of brood rearing. The colonies can only stop breeding again after three weeks, because once started, the brood must be kept at 35°C for three weeks, even if it is very cold during that period. Therefore, the colonies must be kept in winter in such a way that they do not start brooding immediately at the first thaw. Do not wrap up honey bee colonies warmly! Wintering above an open mesh bottom box: "Warm head and cold feet, is the motto in the winter". It does not harm the colonies either if they are not only overwintered "cold from below" (above an open mesh bottom box), but also "cold from above" (with an empty brood box placed on top). This variant arose during the development of varroa control in late autumn/early winter with repeated evaporation of oxalic acid with a stove and a tea light in an empty brood box on top. A procedure how to do this in a very easy, money saving and comfortable way will be dealt with in more detail later on in this book. This evaporation treatment (done in your absence not in your presence at your apiary) is tolerated much better by the colonies than the use of oxalic acid that is sprayed or dripped and therefore this evaporation treatment can be repeated if necessary without harming a colony. The empty brood box will only be removed once it has been established that further treatment is no longer necessary. By counting the fallen varroa mites lying on the inspection board at some point in the winter; each fallen varroa mite in the winter means that there are still 1 x 500 mites in the colony. Varroa control must result in the colony overwintering with no more than 1,000 mites. It has also been investigated how the colonies overwinter when an empty brood box remains on top of the colonies until spring. The result was that overwintering with an empty brood box on top and closed with an inner lid and a (tin) roof (lid) did not harm the colonies. This form of overwintering was compared by Dr. Liebig with the tried and tested variant "cold feet and warm head" in the winters of 2012/2013 to 2018/2019. However, the test in Siberian cold winters is still missing before the phrase "Warm head and cold feet is the motto in winter" can be replaced by "cold feet and a cold head fill the honey pot in spring" ☺.

The care of the colonies during the year

The development of the population is characterised by the rapid strengthening of the populations in the spring, which allows for optimal use of the flower crop (flow) possible, but also essential for the propagation and thus the maintenance of the species and by shrinking the colonies in late summer/autumn for wintering strength with the simultaneous appearance of the winter bees.

All processes taking place in the colony are aimed at fulfilling its genetic and environmentally imposed needs and are effectively attuned to the weather and the prevention of nectar flow. The bee colony is not dependent on interventions by a beekeeper. As a rule, these constitute a disturbance to which the colony always reacts in the same way: it tries to remedy the disturbance and restore the original harmonious state.

The influence of the beekeeper is essentially limited to:

- providing the colony with suitable housing
- possibly ensuring more forage by travelling
- preventing swarming
- replacing the removed honey supplies by feeding sugar water, syrup or dough
- protecting the colony from damage by diseases. In particular, comb hygiene and varroa control are important.

In addition, a beekeeper is involved in the breeding. Gentleness, swarming desire and productivity (harvesting instinct) are the focus here, and for some years now, susceptibility to varroa infestation has also been the focus.

Breeding for performance is oriented towards the fact that stronger colonies produce more honey than weaker ones and, if successful, should lead to the colonies becoming increasingly stronger.

It is questionable whether this actually succeeds or has already succeeded.

An increase in honey production over several years can be genetically determined, but also by the environment. It is extremely difficult to separate the genetically determined performance of the honey bee from that determined by the environment, also because the environment has changed continuously during the time that beekeeping has been practised consistently and is still changing today.

Not only in breeding but also in beekeeping many beekeepers have set themselves the goal of obtaining "more honey from more bees". Anyone who reads manuals and monthly instructions in trade journals and searches the internet will come across a large number of mostly well-founded advice on what to do with the honey bee colonies in order to have performance-capable honey bee colonies during the period of nectar flow. Particularly popular was the so-called stimulation of the colonies in late summer and spring. The colonies had to brood more so that they would become stronger than if they had done so without stimulation.

The thorough scientific analysis of the various stimulus measures has shown that they are unnecessary, because the intended effect does not occur. Not only are they completely unnecessary, without such measures it is also easier. Smart and easy beekeeping!

Other recommendations for beekeeping, especially when they are advertised with the requirements of "true to nature", "species-appropriate", or "ecological", should also be critically questioned. This particularly concerns the "hive problem" with which comb construction is closely linked.

In "true-to-life" beekeeping, the fact is ignored that with the help of frames, foundations and queen excluders, it is much easier to monitor bee health than it is possible with colonies that are kept in "beehives" or on "one box" with "natural combs" or in "batch houses" and with an "adapted brood space".

The same applies to the assessment of swarming. Instead of preventing swarming, anticipating swarming is promoted or even recommending letting swarm, the latter with the message that swarming is linked to the renewal of the combs.

It is always forgotten that this only applies to the swarm that has flown away. The swarmed away colony continues to live on the old combs.

"Swarm anticipation" is a form of swarm prevention in which no attention is paid to whether all bees that are swept or shaken from the combs to form an artificial swarm actually want to swarm.

Equally unfounded are the objections to feeding the colonies with sugar water, syrup or feed dough and to the use of organic acids to combat varroa.

The bee year, the beekeeper's business year, can be divided into five phases, just like the development of the honey bee population: beekeeping in the spring, beekeeping during the swarming season and beekeeping in the summer, late summer and autumn care, hibernation.

As we go through the bee year, we orient ourselves after beginning to be a beekeeper to the fact that it begins and ends practically for the beekeeper with the late summer and autumn care. The preparation of the bee colonies for the winter in the late summer and autumn will become the "alpha and omega" of beekeeping!

The goal of the late summer and autumn care is to winter each colony strong, healthy and provided with a sufficient supply of food. Then the rest of the bee year will proceed naturally. After a successful hibernation, one only has to make sure that no swarms leave. But one can also lose bee colonies through robbery.

The measures to be taken during the year depend on each other and differ in detail from colony to colony and from position to position; because each colony is different and there are always differences between the apiaries. In addition, each year is different. The beekeeper also changes while he/she or as he/she learns.

The beekeeper learns when he/she makes notes during the year, evaluates and saves the measures in the beekeeping. We live in the computer age. So, we should use this facility, especially to store the data collected in the so-called hive cards (see figure below).

Overview of honey bee colony care throughout the year

		care for the colony		
late summer and autumn care	in wintering and hibernation	in the spring	during the swarming period	in the summer
end of July till September	October till February	March/April	April till June	June and July
mite fall diagnosis	mite fall diagnosis	check food supply	enlarge	honey harvest
divide overage colonies and treat with oxalic acid.	unite weak colonies	equalise populations strengths	cut drone brood	care for the young colonies
reduce overage colonies and treat with formic acid.	and replace in overage colonies the queen	enlarge	bloodletting and population propagation through queen breeding	
treat young colonies with formic acid	remove remaining mites as much as possible with oxalic acid	remove winter food, give foundations or freshly built combs	treat brood free splits with oxalic acid	mite fall diagnosis
feed overage colonies, "splits" and young colonies	melt old combs		regular swarm checks and swarm prevention	

In the normal keeping of colonies, about thirty population checks (in April-June every seven days) are part of the year. In these, notes on the condition and behaviour of the colonies are recorded or marked, because usually one knows what jobs have to be done.

Hive card

The hive cards are prepared in such a way that the result of an assessment of a colony is quickly recorded. At home, the notes are typed preferably into an Excel file in which the progress of the "bee year" is stored colony by colony, which makes it possible to compare colonies, apiaries and years possible. How the hive cards can look per month is brought forward in the description of the care for the colonies from January to December.

Hive Nr.:
Date:
Time:
Bee flights/Minute:

Brood box above (number):										
Comb nr.	Bees	Work brood sealed	Work brood larvae	Work brood eggs	Drones	Drone brood closed	Drone brood open	Pollen	Honey/ Feed (gr)	Remarks
Wall										
1a										
1b										
2a										
2b										
3a										
3b										
4a										
4b										
5a										
5b										
6a										
6b										
7a										
7b										
8a										
8b										
9a										
9b										
10a										
10b										
Total										

Brood box under (number):										
Comb nr.	Bees	Work brood sealed	Work brood larvae	Work brood eggs	Drones	Drone brood closed	Drone brood open	Pollen	Honey/ Feed (gr)	Remarks
Wall										
1a										
1b										
2a										
2b										
3a										
3b										
4a										
4b										
5a										
5b										
6a										
6b										
7a										
7b										
8a										
8b										
9a										
9b										
10a										
10b										
Total										

Remarks:

Queen seen?
Swarm mood? (Queen cells?)
Gentleness? (1-5)
Disease symptomes (nosema, chalkbrood, etc.)?

Has there been made a split?
Honey box ? (number of combs ?)
Varroamites counted?
Anti-varroamite treatment?

Honey harvest
Date:
.... kg

A hive card (for noting gathered data during a hive inspection)

Do avoid robbery

It happens time and again, especially in the nectar-poor summer season, but also in the fall and spring, that colonies are robbed of their food by other colonies. In most cases, the beekeeper is to blame; because every robbery is preceded by stupidity. That is why it is important to avoid everything that provokes or promotes robbery at all times:

- the entrances of splits and weak colonies should be kept small in principle
- any unnecessary disturbance and keeping the colonies open for too long should be avoided
- feeding is only done in the evening and given in reliable, leak-proof feeders
- it is best to feed "from above" and not to spill sugar water or syrup
- do not leave combs, feeders or feeding barrels open, not even to let them "drain"
- do not let the colony die from varroa infestation

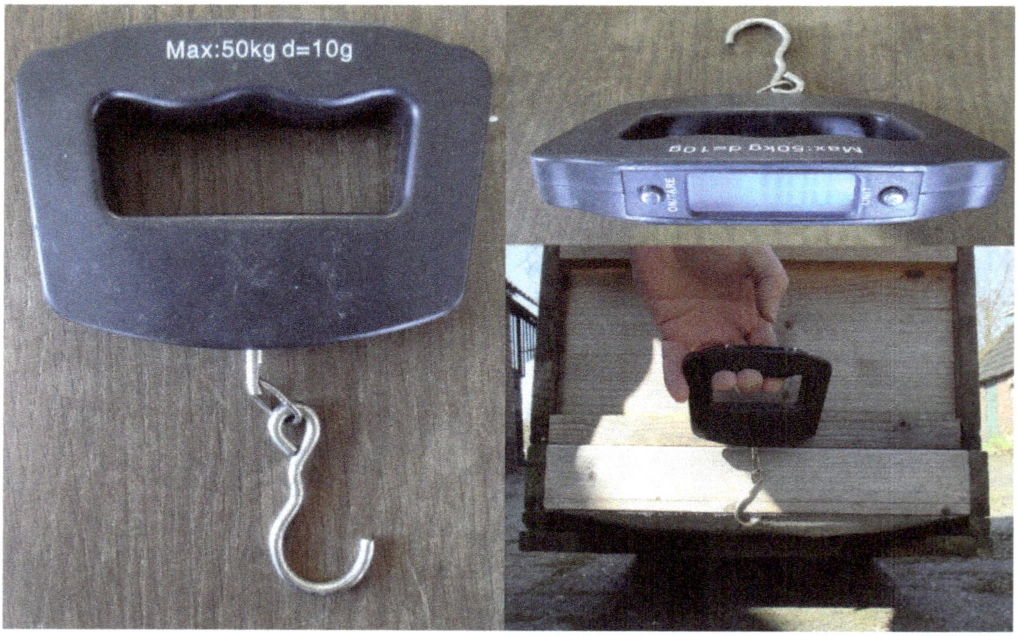

A digital suitcase scale. This device that can be obtained for little money enables to determine the easy way the weight of a colony/a colony's food supplies. It is operated "from the back". For knowing the total weight the indicated number of kg has to be multiplied with two.

The care of the colonies in spring and early summer

The wintering out

Usually in March the days in which the weather makes a cleansing flight or even a gestation flight possible for the honey bees increase. These are opportunities for the beekeeper to inform himself about the condition of his colonies. It is sufficient to limit oneself to observations of flight behaviour and the control of food supplies.

One will then act then in the same way as during the final inspection in October.

Each colony is lifted up a little and arranged in this way from the heaviest to the lightest. Or the colonies are weighed using a digital suitcase scale "from the back". The result is noted in the apiary protocol. If you are not sure which food supply corresponds to the "full" or "half" weight, then the lightest colony is opened. After the bees have been driven down the seams (the spaces between the combs) with some smoke, a glance from above is sufficient to see the size of the food sources. The following applies: as long as sealed feeding cells are available or visible above or/and behind (as seen from the entrance) the seat of the bee cluster, the colony will not suffer from a lack of food. But this can change overnight in the spring when food supplies are low.

Anyone who does not want to be satisfied with just looking under the foil and wants to know more precisely can look at each frame of the top breeding box and then estimate the food supply. A assessment frame or the "eighths method" comes in handy when it is determined how many eighths of the frame surface on each side of the frame are still filled with food.

One eighth (Zander frame) corresponds to one dm^2 frame area. There are approximately 400 worker brood cells on a dm^2 frame surface. When 400 cells are filled with food, they contain approximately 125 grams of food. Sixteen eighths correspond to a Zander frame filled on both sides and therefore approximately 2 kg.

The (Swiss) ***Liebfelder Estimation Method****, Each unity ('square') of a frame in Zander format represents either 125 bees, or 400 cells with worker brood or food (=125 grams of food) or 230 drone cells.*

This way you can assess easily and with high accuracy population strength or development or the food supplies in spring or autumn.

So, one side of a Zander frame: 3,200 cells, ca. 1,000 honey bees, ca. 1 kg food and about 1,840 drone cells.

Pictures after Dr. P. Aumeier

Frame size	*Bees*	*Drones*	*Worker brood*	*Drone brood*	*Pollen*	*Food/ Honey*
Zander	*125*	*100*	*400*	*230*	*40 g*	*125 g*
BNHF	*110*	*90*	*350*	*200*	*34 g*	*110 g*
Langstroth	*140*	*112*	*450*	*259*	*45 g*	*140 g*
Dadant	*176*	*141*	*564*	*325*	*56 g*	*176 g*

A comb with food (honey and pollen) and a small brood nest.
Every unity or 'square' of the Liebfelder Estimation contains 125 grams of food.
How much food is there on the comb in this picture?
And how much sealed brood cells?

Pictures after Dr. P. Aumeier

It is sufficient to estimate the food supply among the lightest colony. On this occasion you can also immediately record the population strength (bee number and brood size) and the pollen supply.

The temptation is then great to also subject other colonies to a population estimate. This is only done regularly to investigate the sense or value of new ideas and new products that, in the age of the Internet, are quickly finding followers among experienced and 'young' beekeepers. However, it should be remembered that any intervention so soon after wintering is an unnecessary disruption. If frames are raised, this could cost the lives of some bees. The queen is also in danger.

Those who have wintered their colonies "smartly" do not have to care for them in March. If mistakes were made during the summer care of the previous year, they cannot be corrected now. Weakly outwintered colonies often cannot meet the special requirements imposed on them by the changeable weather. They are more susceptible to diseases, do not progress well and even die. The beekeeper is usually to blame. it doesn't help to complain about the supposed bad weather. Changeable weather in March and April is normal.

It is one of the characteristics of a bee colony that it is always able to withstand the vagaries of the weather if it is healthy and has sufficient food supplies.

The assessment frame as explained. An estimation frame is made quickly and easily with four elastic bands.

After a while, with some experience, it becomes possible to execute the estimation with 'virtual' units as well.

The spring inspection

After the cleansing flight, a warm day when the bees are flying is used for a thorough spring inspection. For each colony the number of fallen in the lower mesh bottom box is examined; a look through the opened entrance with the aid of a flashlight is sufficient. If there are many dead bees on the bottom - usually they have collected near the entrance - and the bottom is very dirty, it should be replaced. This is especially necessary if the fallen dead honey bees in winter are wet and moldy, which is especially common in closed hive bodies and in plastic hive bodies.

In the case of colonies that have overwintered above an open mesh bottom box, this work can generally be avoided. If dead bees have accumulated and you want to save the bees the work of cleaning them up, it is sufficient to remove the mouse mesh, tilt the colony forward and hit the mesh from below until the dry bee corpses with the coarse material lying on the bottom, having fallen out of the entrance. The remaining cleaning work can be left to the bees.

You can also remove the colony sitting on one or two breeding boxes from the bottom box and park them on the tin roof placed behind it. The fallen on the mesh bottom box can then be shaken into a bucket for further examination. After a quick cleaning, the mesh bottom box is returned to its old place and the colony is placed on top of it again. One can execute the practice of "picking up and putting aside", "cleaning", "setting up again", even during heavy flying activity that is assessed immediately prior to this. It is recommended to estimate the number of fallen of each colony, either with the "eighths method" when the dead bees are still in the mesh bottom box or one pours the fallen ones into a measuring cup or glass honey jar. 1,000 bees fill a one-pound glass honey jar.

The collected debris from fallen victims can also be examined for fallen varroa mites.

To this end, everything that has fallen is placed in a double sieve and showered under the tap. The mites all end up in the fine sieve and can then be counted. This washing can also be practised in May/June when examining the sealed drone brood that serves as a "varroa trap" during the swarming period and is cut from the frame before the drones emerge and so do the varroa mites. The values found are noted in the stock protocol and later compared with the further development of the colony.

The tilt control (again to make life easy as well)

From the beginning of May on, the bee colonies must be checked every seven days to see whether they are in swarming mood. The seven-day rhythm is also reflected in the mnemonic "three, five, eight and the queen is made". It reminds us of the development period of sixteen days of a queen. Three days egg, five days larva and eight days pupa. The seven-day rhythm must be strictly adhered to during the swarming period so that the swarming mood is kept in check and controlled at the apiary. Swarming must be prevented.

The swarming period, for that matter, ends somewhat after the longest day.

During the tilt check, the two breeding boxes are separated by tilting the top box upwards and at the same time pulling it slightly forward. This makes it easy to see whether swarm cells (queen cells) have been created in the bee colony. If that is the case, they must of course be removed. If necessary, a flashlight can be used for a better look.

Swarm cells are easy to recognise and they are usually built at the bottom of the frames. If there are swarm cells, each frame must be checked. If no swarm cells are visible during the tilt check, then, as years of experience with hundreds of colonies have shown, there is hardly any chance that there are swarm cells in the bee colony.

Thus, the tilt control helps make beekeeping easier and less time-consuming. It is a 95% reliable swarm control method.

A tilt control carried out in winter. The honey bees are huddled together in a ball (cluster) that is now divided in half by the tilt control. The honey bee cluster is close to the entrance.

Instructions for cleaning the mesh bottom box in the spring

During air traffic or even after flying:

- remove the metal lid and place it upside down behind the colony on an empty brood box
- use a hive tool to loosen the lower breeding box from the bottom box
- remove the colony from the bottom box, give it some smoke at beforehand if necessary, and place it on the inverted metal roof
- clean the mesh base and collect the dead bees in a bucket - put the colony back on the bottom box

Each colony is also checked for food supply, strength and queen rightness (presence of the queen). There is no need to take any frame during this check.

As already described, the food supply can be estimated by lifting the colony. The strength is assessed by means of the occupied seams (the spaces between the combs) and air traffic. Which colony fly more, which fly less? Take a clock with a second hand and try to note how many bees return to the hive per minute for each colony. There could easily be more than a hundred! Do not open the hives in advance, because any disturbance will cause the colonies to stop flying or at least severely limit it.

To determine whether a queen is present, the presence of worker brood is sufficient.

One should not look for the queen or look at how many eggs she has laid in the past few days. It is sufficient that sealed worker brood is seen. It is not necessary to raise frames for this. When lifting and replacing the frames, there is always the risk that bees will be crushed or even crushed to death. If that happens to the queen, one can no longer count on a harvest of spring honey. One should not want to take such a risk. It is sufficient to drive the honey bees downwards with a few strong puffs of smoke, so that a view can be taken of the breeding surfaces. If no brood can be detected in this way and there are also other indications that the population is not doing well (many dead bees, relatively large food supplies, few comb seams populated with honey bees, little air traffic, relatively few pollen collectors), a further check of the colony in which several frames are lifted will not be detrimental to accurately determine the state of affairs of the colony. With all other colonies that look good, it is better to omit this.

Weak colonies do not need to be cleared if there is still a queen present and they are not demonstrably suffering from a disease. In every healthy bee colony, even if only three comb seams are occupied in March, there is an enormous growth potential of colonies that are so weak in the spring, although as a rule no large honey harvest can be produced from the flow in spring, but they can be so strong at the end of spring that they will still perform well for the honey harvest in the summer, provided there is sufficient flow of course. If summer flow does not occur, such colonies will only have incurred costs in their formation in June, in care during the summer, in staging and varroa treatment in late summer and autumn. To prevent something like this from happening too often, you should ask yourself what the cause of poor wintering is. Then all notes will come in handy, as will exchanges of experience with other beekeepers.

In the spring (as in late summer) robbery can easily break out. therefore, the entrance must be kept small in weak colonies.

Boosting weak colonies in spring

Scientific research into "double population management" has produced a method by which a weak population, which only occupies about 1-3 seams in March, can be "helped on its way" without filling them. It is placed on top of a queen excluder on a fairly strong colony and then removed after 4 to 5 weeks. During this period both populations have a common entrance.

The top colony receives care support from below and it grows in no time with a breeding nest of more than six to eight combs. But of course only if there is enough room for it! Therefore, in the case of weak colonies, the redundant full feeding frames must be replaced by empty frames before remediation takes place. Two filled edge frames are sufficient for the colony as a food supply. As soon as the first brood raised by the foreign honey bees (!) emerges, the brood box can be removed and placed (again) on its own mesh bottom box. By then it is also the time that the cherry trees blossom. If you prefer to be certain to keep all the flying bees in the right hive, so-to-say, you can put this hive at a location outside the flight range (> 2 km).

The periodic combination of a strong colony with a weak colony is associated with a slight reduction in nurse bees, but this does not hinder its development. Both colonies can be increased in size after separation.

The bottom colony gets the honey box on top of the queen excluder. First, an edge frame is exchanged for a building frame (see p. 109), which is hung in position 2 or 9 in the breeding box.

The remediated, relocated population is simultaneously enlarged upwards (above the queen excluder) and downwards. Both new boxes will be equipped with a frame core of four ready built frames and 2x3 foundation frames. If the breeding nest of the remediated colony contains more than eight windows, the first building frame can be hung between two ready built frames in the lower breeding chamber. You can also hang a breeding frame downwards to create space for a building frame in the upper breeding box.

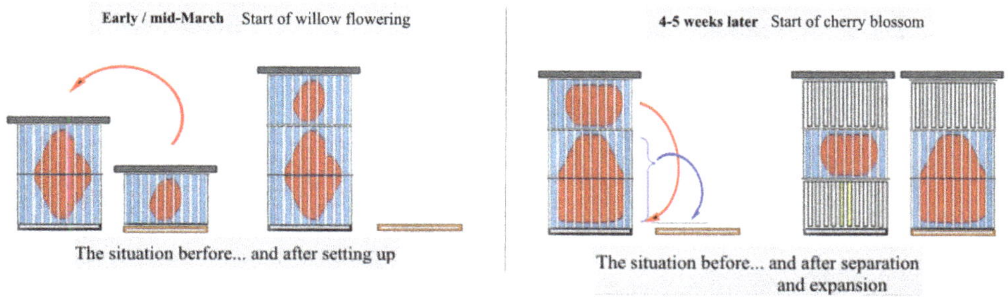

The remediation of a weak colony in early spring. This includes a period of 4 to 5 weeks. After separation, both colonies can immediately be enlarged. You can also wait until the swarming period has started with the separation, but then the colonies will also need a communal honey chamber above a second queen excluder before then. A fly opening (more an exit than an entrance) for the drones must than also be made in the breeding area of the upper colony.

The separation of the double colony can also be implemented in such a way that both colonies remain in the same position. Then the bottom colony is moved and the top colony then take the place where the bottom colony used to be. In this way it receives all the flying bees from the double colony, which then makes it necessary to increase the size of one in three boxes (see figure).

Emergency feeding

If the colonies have been wintered in two breeding boxes and have been adequately fed, then in March there will still be a wide feeding edge above the breeding nest and food supply is guaranteed. This can quickly change during the spring, especially in changeable weather. During periods of cold, the colonies continue to breed, but then need to warm up more. In unfavourable weather conditions, a colony can become two kg lighter within two weeks.

Especially in persistently cool and rainy weather in March and April, the food supply must be checked again and again. one is only freed from this worry when the great flowering begins and the colony can use fruit and meadow flowers.

The result of each food supply check is recorded. The light colonies are usually also the strongest; for they breed more strongly and therefore consume more food than the colonies who are weaker in this respect. If a colony is found too lightly, ensure that it needs to be provided with a feeding frame, which can be obtained in advance from a colony better equipped with food (or that is in reserve from the honey harvest of the previous year and was not thrown away). Such a feeding frame must be hygienically clean, without traces of droppings.

The spread of diseases can happen through the exchange of frames, so it is better if one can refrain from doing so. If the colonies have been fed sufficiently in late summer, it is not necessary at all.

It is wrong in principle to feed the colonies extra in the spring. Feeding with sugar water, feed dough or syrup just before the flow always threatens the purity of the nectar and the honey made from it. When the food supply becomes too low for all or almost all colonies in the spring, the survival of the colonies is at risk if the feeding pots are not reached. The lesson to be learned from such an experience is 'Once, never again!' It is better to give more food in late summer and remove excess winter food in spring.

Honey is most suitable as a supplementary feed in case of emergency in the spring. Obviously own unsuspected honey can be given. Hands off of honey that is not your own. With such honey there is always the risk that it contains traces of the pathogens of the evil foul brood. Feeding not own honey can infect a colony. If it then becomes ill, the disease can easily be transmitted to nearby bee colonies. The damage would be enormous and a definitive cull would be a lot of hassle.

If there is no unsuspected honey and no clean feeding frames available, feed dough can be used, because it can be given directly to the colony without the use of further aids. Under the foil, a thin slice is simply placed on the colony and covered with the concave side downwards of the turned over cover plate.

Small portions of up to one kg are sufficient and the administration is repeated if the flow is delayed.

An alternative to feeding "feed dough from above" is feeding with "feed syrup or sugar water from below". A flat feeding bowl is placed on the mesh bottom, filled with syrup or sugar water and swimming aids (corks, twigs, leafs, etc.) are placed on top. The entrance is reduced in advance with strips of foam rubber.

What to do with a queenless colony?

Queenless colonies can no longer be saved in the spring. The queen was probably lost during an intervention the previous autumn. Either the affected colony was able to create a young queen when the youngest brood was available and spent the winter with an unfertilised one, or it could not have created a new queen because it was brood-free. Then some workers can and do become "drone mothers" and lay eggs. The eggs of the laying workers (called "drone mothers" on the continent) are not "stuck" to the back wall of the cell, because their abdomen is not long enough for that. The eggs are "glued" to the side wall. Often several eggs are laid in a cell.

In both cases the colony will have only drone brood in March. Then it is lost and most certainly not suitable to be put above another colony to strengthen it. This brings only the queen of the intact colony in danger. Dissolving a drone breeding colony at the wrong time for the purpose of strengthening another weak colony can easily lead to the loss of that colony's queen. That is why you can find clarity before dissolution by means of a "queen test". To this end, a breeding frame of a queen right colony is placed which has a few open cells with youngest brood and/or a few cells with eggs underneath, in the presumably queenless colony, after which a check is made after a few days to see whether the colony is in the process of "creating" a new queen.

If the result is positive, the colony can be left to themselves for the time being. The colony is dissolved before the new queens hatch. This is best done when the honey bees are flying. The bees are shaken in front of a previously reduced entrance of another colony, which is also given a "queen test" after breaking the mother caps. If no rescue cells are formed on that given brood frame, the queen can be sought. She is removed, after which the colony is dissolved.

It is easier, for that matter, to dissolve a drone brood colony, as confirmed after a queen test, some forty meters or more away from the apiary by shaking off the frames. The drone mothers are unable to bridge the distance and harm an intact colony.

Set them wider in the spring

The first brood raised in February and March is usually sufficient to compensate for the loss of winter bees. As temperatures rise, the breeding nest is further expanded. The result is: the colonies grow, at first slowly but soon faster.

The pace and extent are determined depending on the weather, the pollen introduced and the strength of the colony during wintering. The equations apply: "more bees =

more brood" and "more brood = more bees".

After a mild winter, the well-cared for colonies start growing in March, as long as there have been consistently mild weather conditions in February and March. The strong population of honey bees can continue until June if the colonies do not get into the habit of swarming. New bees emerge every day, more in a healthy colony than there will be bees that die of old age. At the beginning of April, about 500 young bees are born every day, and by the end of April even about 1,000 or more, every day.

The bees need more space than the brood. Each side of a fully developed Zander frame contains approximately 3,200 worker cells and can house approximately 1,000 honey bees when it is very densely occupied. That is rarely the case. A seam of bees is considered well filled if there are 1,000 honey bees on both sides of the frame. A weekly increase of 7,000 to 10,000 young honey bees can easily fill seven to ten comb seams. Then a honey bee colony can quickly become (too) small, especially if it lives in one brood box, but even as well if it has overwintered in two breeding boxes.

One-box-colonies are usually enlarged from "one to two" during the flowering of the willow in March, two-chamber colonies from "two to three" when the cherry trees bloom in April. The third box is given as a honey room. (So there is a queen excluder between the second and third chamber). If the flow is good, a next box will follow in May during rapeseed flowering. For each larger installation, four to six ready built frames and six to four foundations are placed respectively. For larger hives with two boxes, eight to twelve extended frames must be prepared for each colony. More will be needed later. When forming and caring for young colonies, another six to eight foundations are needed for each split. Moreover, one must always be prepared for swarms to arrive. In principle, each swarm is placed exclusively on foundations. One then needs ten of these for each swarm.

The foundations must be pressed and secured in the frames before work outside on the colonies begins in the spring. Anyone who started beekeeping in the spring and does not yet have ready built frames will of course only use foundation frames to enlarge the colonies. There is then no other choice.

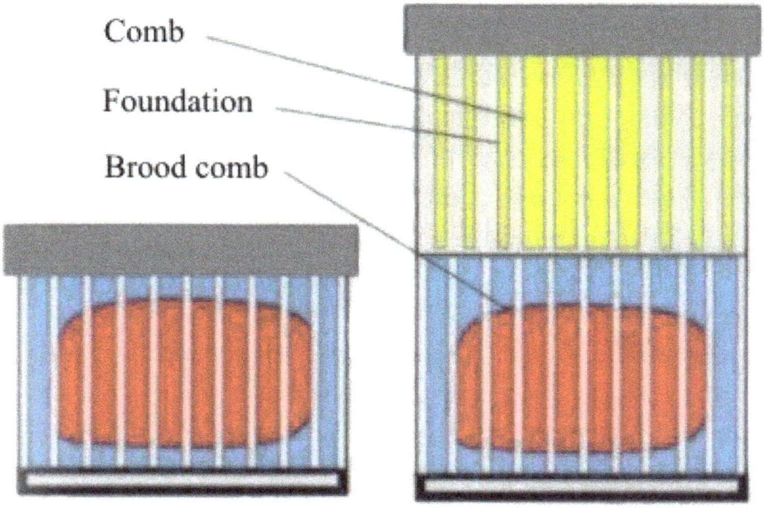

Situation of a colony wintered in a room before and after enlargement.

The building frame.

The building- or construction frame is an empty frame that is mainly used in April, May and June. If it is hung in the brood nest or on the edge of it, the colonies usually build only drone cells (comb) that are immediately occupied by the queen with unfertilised eggs. In strong colonies it only takes a few days before the building frame is completely built and provided with eggs, so that the drone brood can be completely sealed already two weeks after giving the building frame. After three weeks this is usually the case in every colony if it has not previously become swarming.

After wiping or shaking off the bees (be careful if the queen is on that frame as she often is situated on the comb with the drone cells), the sealed drone brood is cut out and the building frame is returned empty into the colony. This process is repeated three to five times every two to three weeks in the spring and early summer. After the last time the building frame has been cut out, it is not returned but replaced by a foundation frame.

The building frame can be useful in several ways:

- It serves as a "swarm lust barometer".
- If the building frame is being expanded expeditiously, it is a sign that "everything is in order". The colony has a good queen - because a queenless colony does not build - and it does not want or want already swarming; because as soon as it gets into the mood of swarming, all building is stopped.

- It dampens the urge to swarm.
- If it (the building frame with drone brood) is cut out or is replaced by an empty building frame the colony will be kept busy with further building and breeding care. As long as a colony builds and can take care of many larvae, it will not (yet) want to swarm.
- It serves as a varroa trap.
- Drone brood is more attractive to varroa mites than worker brood (due to its longer development time of drones], moreover, varroa breed more numerously in drone brood. When the sealed drone brood is removed, many varroa mites are also removed, which demonstrably slows down the degree of varroa infestation in the population. If the drones that have grown up in the building frame emerge, exactly the opposite is achieved.
- The cut drone combs yield clean wax after melting.
- The foundations given by expanding are build more beautifully and with less drone cells interwoven.

In principle, the first building frame is only provided when setting up the honey box. This also applies to the colonies that have been overwintered in one box hive.
There are two methods for enlargerging one-box-colonies in the spring:

1. The stronger one-box-colonies are given the second box already during the willow flowering and that is provided with a frame core consisting of four-six already built frames. On both sides of two or three foundations are hung in the frame core. During cherry blossom the setting up the honey box follows that is linked to the placinging of building frames in the top breeding box at position two or nine. The already built frame out of the foundation in that location - which is usually also already filled with (some) honey - is given a place in the honey box.
2. The weaker single-box-colonies only will be enlarged shortly before or during cherry blossom, but also with two boxes at the same time. An extra breeding box is provided with an already built frame core and foundations (or only with foundation frames if one has no already built frames [yet]) is placed under the existing breeding box, the second box is placed as a honey chamber on top of a queen excluder and on the wintering box. First an edge frame is taken out from that breeding box and the next frame is pushed towards the wall of the box. In the resulting empty space a building frame hung. The removed edge frame will be the best, especially if it is also has food or pollen, placed in the middle of the lower box.

The entrance of the colony, enlarged from below, remains or is reduced. (Maximum size for that matter, of a entrance of a honey bee nest, which bees themselves seem to have a preference for is 70 x 40 mm) Simultaneously enlarging one-box colonies from below and on top ("from one in three") ensures better acceptance of the honey chamber than with an increase gradually from one to two and later from two to three. In a warm spring vegetation grows faster than honey bee colonies. The "big bloom" then starts before the (young) colonies overwintered in a one box occupy the second breeding box completely build up and filled with brood. The second brood box is often (also) used by the honey bees at an early start of the flow as a honey chamber, the brood nest stored in it is pressed down by the honey stored in it and only hesitantly the honey box is accepted that was set up later.

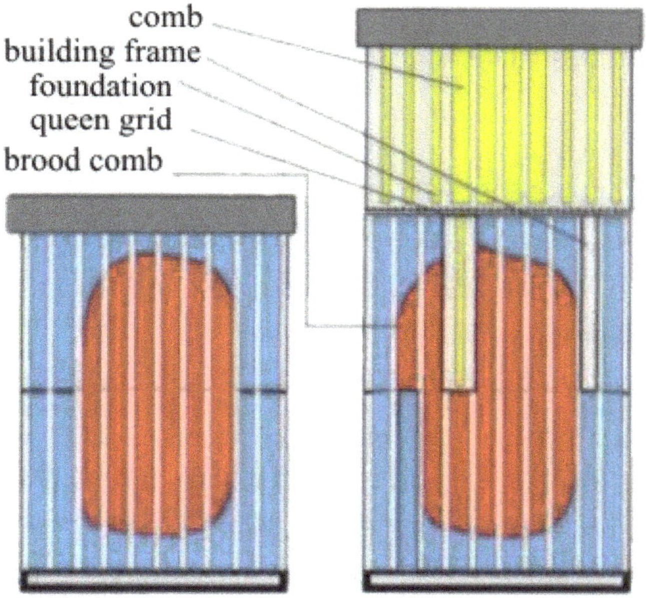

Situation of a colony overwintered in two boxes (left) after placing the building frame (construction frame) and a foundation frame in the upper brood nest that is thus stretched out. The honey box (the super) is then placed on top of the queen excluder (right). Two edge frames have been removed.

Note again that all boxes have the same size.

Then you will find more honey under the queen excluder than above in the honey box, which is often necessitates to remove the queen excluder. This will prove itself later with the honey harvest as a wrong measure.

The method of increasing one-box-colonies from "one to three"

- prepare two boxes for enlarging (U ‹= to place underneath› and H ‹as a honey chamber›), equipped with either four already built frames and foundations to the right and left of them or just all foundation frames. The upper brood box will have a building frame.
- open the hive of the colony, remove an edge frame and hang it with bees in the middle of the enlarging U-box
- slide the frame next to the lifted frame towards the wall
- hang the building frame on the resulting empty spot (in position two or nine).
- pick up the one-box colony and place them on its side in the front above or place it on the inverted tin roof. Place the box U on the mesh bottom box.
- Return the colony up box U, remove any wild built comb at the bottom beforehand and collect that without bees in a bucket and/or expel a cluster of bees that are sagging from below with a few puffs of smoke between the seams.
- place the queen excluder
- place box H on top
- cover the colony with foil, inner lid and roof lid
- if necessary, reduce the entrance

The building frame can also be hung in the upper breeding chamber in the middle of the breeding nest. But then there is the risk that, in addition to drone cells, worker cells are also built on them, making it more difficult to cut out the sealed drone cells later.

Weaker colonies in particular tend towards "mixed farming". That is why they should also place the building frame in position two or nine, even if their breeding nest has not yet expanded to frame three or eight. On the other hand, it makes little sense to hang the building frame in the lower breeding box. If it can then be cut out, the second box must first be lifted and set aside. You can save yourself that work.

In any case, the drone brood must be removed before the drones hatch if it is to be used as a varroa trap. Also you don't want extra varroa mites in the colony.

The use of a building frame requires consistent periodic work. During the swarming season in May and June, cutting is linked to regular swarming checks. It is advantageous if the production colonies at an apiary are cared for synchronously. This sets limits on the hustle and bustle of the associated side activities (melting drone brood cells and, if necessary, examining them in advance for the degree of varroa contamination).

Setting up the honey box (the honey super)

Strong colonies are ready for a honey super at the beginning of the cherry blossom. This can be the case between the beginning and middle of April, depending on the weather and the location (climate). In general, the rule applies to every upcoming larger setting: There is no "too early", only "too late"!

The boxes for the larger setting are always prepared at home. In the middle of each box for the wider setting, four to six built-out frames are placed and next to them on the right and left three or two foundations. If there are no built-out frames available, the setting can only be done with foundations!

Before the honey super is set up, the food supply and colony strength are assessed. To this end, the upper brood box is tilted upwards and the frames are examined from below to determine which frames are brooded, whether the colony is in the middle or on one side and how well the frames on the left or right are filled with food. Often the result of this assessment corresponds with the impression previously obtained when observing the air traffic at the entrance

After this "view from below" the upper brood box is put back, one of the two edge frames is lifted, the frame next to it is pushed to the wall and a building frame is hung in the empty spot. Then the queen excluder is placed on top - if necessary the superstructure is scraped off the top bars of the frames beforehand and collected in a ready bucket - and the prepared honey box is placed on top. Then the colony (the beehive) is closed with foil, inner lid and roof lid. Which edge frame is "harvested"? For a few years now beekeepers have been using a feeding trough when feeding the honey bees in late summer that is hung in the upper box on the edge and that is filled two to three times with syrup (or sugar water) before feeding. After the supplementary feeding is finished, that feeding trough is removed. Two extended frames or foundations are hung in the freed up spot. These are usually still empty in the spring.

If the colony is relatively light at the beginning of April, one of these two empty edge frames is removed. If the colony is relatively heavy, a full frame is "harvested" on the other side and stored insect-proof in a cool place until the end of April / beginning of May. These feeding frames are then used for the splits that are formed around that time.

What I don't do when increasing the size!

- I don't hang brood frames in the honey chamber
- I don't do floating feeding

Floating feeding or other stimulation measures used to be very popular with continental beekeepers. These days, these activities are being rediscovered by many authors....

Stimulation could increase the brooding activity of the colonies. The research in the Hohenheim Honey Bee Research Institute (in which Dr. Gerhard Liebig has worked) has clearly proven that it is not possible to induce the honey bee colonies to brood more than they want or can by any artificial intervention. Both the repeated scratching open of sealed food and the hanging in of a feeding trough filled with feed dough or sugared honey, as well as repeated liquid feeding with, for example, one litre of sugar water (1:1) at intervals of two or three days are unnecessary measures that only cost effort and money and have absolutely no positive effect.

- For the same reason you also should not swap the brood chambers (top to bottom, bottom to top) and
- also should not turn the top brood box 180 degrees.

These incentive measures are described and recommended in much professional literature. Their supposed effect is substantiated by the fact that the bees are forced to occupy themselves with the food or to restore order when disturbed. During their cleaning activities, the bees produce -in the imagination of the authors- extra heat, an intensified food flow and free more cells, which should then induce the queen to lay more eggs. This substantiation sounds logical but is nevertheless nonsense. The colonies do not allow themselves to be stimulated!

In the event that more than one feeding frame can be harvested - which rarely happens - the colony in question not only gets a building frame (in place two or nine), but also one or two foundations. These foundations are placed "in the middle of the brood nest" or between two brood frames. There, a foundation is built out very quickly and immediately occupied with eggs by the queen after the expansion is finished.

With this method, the brood nest is pulled in width in the upper box, which makes it easier to accept the honey super that is placed next to the queen excluder.

After setting up the honey super, all colonies of an apiary are synchronously cared for with regard to the building frame. At the beginning of the swarming period (towards the end of April/beginning of May), each bee colony can also work with two building frames, which are "harvested" alternately after the sealing of the drone brood. This removal of drone brood is done weekly (= at intervals of seven days) in May and June during the swarming mood checks.

After the colony has been arranged in the upper brood chamber, the food supply should still amount to at least three kg. This is available if there are two or three finger-wide feeding wreaths on the brooded frames in the upper box above the brood. The beekeeper must ensure that during beekeeping interventions, no drones, no drone brood and most certainly also not the queen end up above the queen excluder in the honey box. They would then be trapped there forever. The queen will not be bothered by this because she will still be well cared for in the honey chamber, but if the honey frames are already well filled there, she cannot fully utilise her egg-laying capacity. The drones will not tolerate being trapped at all.

Don't let them swarm!

A swarmed population no longer produces honey. Even if the beekeeper manages to catch the swarm before it travels far and wide and houses it in a beehive, the swarm is not suitable for honey extraction. Until the first brood hatches and the colony begins to grow again, fewer than 10,000 of the 20,000 bees the beekeeper has housed remain. And honey cannot be made with that, even if there is a good flow. Only strong colonies produce honey and these must contain at least 30,000 bees during the flow period. This strength is no longer achieved by the swarmed colony, even if it does not lose any subsequent swarms and the young queen quickly starts laying. Till the first brood emerges, the number of bees decreases enormously. If bee colonies have to supply honey, they are not allowed to swarm. Anyone who leaves his colony to themselves during the swarm period takes that risk. Swarm beekeeping is opposed to swarm-free beekeeping. Swarm-free beekeeping works to prevent swarming and preventing swarms. By taking measures to prevent swarming, the beekeeper tries to ensure that a colony as a whole will not get into a swarming mood. If this does happen, he will take measures to prevent swarming

Early swarming was recommended in old beekeeping manuals from the time of straw hive beekeeping. At that time, bee colonies were housed stably (for example in straw hives) and kept in such a way that they swarmed. The swarms were captured and used for honey production. In order to harvest the honey, the colonies were sulfurised and the combs were cut out. The earlier the swarm went off, the more honey it could produce. With the introduction of the "mobile house" beekeeping changed. The bees are kept on movable frames in hives that are accessible from the back or above. With this turn in beekeeping, it was discovered over a hundred years ago that colonies produce more honey if they do not swarm.

A swarm at the table. One might wonder whether the honeycomb pattern of the stones under the table exerted a certain attraction...

Preventing swarms

The measures to prevent swarming can be summarised from the objectives in a simple formula: "Avoid unemployment!" Above all, the measures must be taken in a timely manner and aim to ensure that the colony always has sufficient opportunity for brood care. The beekeeper can achieve this if he:

- regularly uses the building (construction) frame
- provides sufficient space
- timely 'bloodletting' (a well filled brood frame is removed from the breeding nest and a foundation hung in its place. This brood frame is used in a split with brood and honey bees to form a new colony.)

Anyone who refrains from this should not be surprised that swarming mood arises.

The building frame given in April is not only used as a varroa trap (and is therefore removed or cut out before the drones emerge); it should also help prevent swarming mood from arising.

From approximately the beginning of May, two building frames can be used simultaneously but at time intervals, so that the bees always have the opportunity to build freely and breed drones. The prompt expansion of the building frame already signals during the tilt inspection that the situation on the labour market is still relaxed. The drone maggots are good food takers.

A further important measure is the timely bloodletting of the colony. Usually one contents with bloodletting the colonies once during the flowering of the fruit trees. This intervention must not be at the expense of the colony's capital and must not affect their production capacity to utilise a possible later summer crop. This is most likely guaranteed if the colonies are bloodletted only once, at the end of April or beginning of May and then be bled very carefully. This means: removing one or at most two brood frames with predominantly sealed worker brood and the bees sitting on it. In addition, the bees can be added that are located on a sealed and therefore 'harvest-ready' drone brood frame that had been hung in the hive as a building frame 21 days before. Check carefully if the queen is not on one of these frames.

A split with brood and bees is made with the honey bee and brood material taken from the colony. From four colonies onwards it is worthwhile to create a collective split with brood and bees that can be used as a foster colony for queen breeding after nine days and that is divided into fertilisation colonies twelve days later, when the young queens emerge. The details of this queen breeding method are described later in the chapter "Colony multiplication in four steps with integrated queen breeding".

With adequate care, the split with brood and bees or the fertilisation colonies usually develop until October into colonies capable of overwintering.

Early bloodletting leads to a smaller reduction in production during the blossom period, but the urge to swarm does not break out as violently. That effect has more weight. If you only bleed when the colonies are in swarming mood, you must take many more bees and brood from the colony to achieve success. Moreover, you permanently weaken your colony. The honey production from a subsequent summer flow is then worse.

It is best to connect the (early) bloodletting with the harvest of the first drone brood frame, which is realised approximately 21 days after the installation of the first building frame. The drone brood must have been sealed.

The size of a collective split with brood and bees depends on the number of colonies that are bloodletted. Weak colonies are not bloodletted!

If four to nine colonies are available, the collective split with brood and bees is made with a maximum of ten brood frames (and a feeding frame) in one (brood) box (with foil, lid and a roof on top and placed on a mesh bottom box). If more colonies need to be bloodletted, one can make a collective split with brood and bees of two boxes or (better) two collective splits with brood and bees in one hive each.

The formation of a split with brood and bees

The procedure of "Colony multiplication in four steps with integrated queen breeding" only pays off with a larger apiary. But also in smaller beekeeping apiaries, great importance must be attached to the rejuvenation of the honey bee populations. Young colonies can be created via the split with bees and brood already described or via a split with bees and a -mostly already fertilised-queen) that one does not have to take further into care (in the beginning).

The later the split is created, the stronger it must be formed if it is to develop until autumn on its own into a colony capable of overwintering. If formed at the end of

April/beginning of May, one brooding frame is sufficient, in mid/end of May two brooding frames are needed, at the beginning of June even three, each time with the bees sitting on them.

The brood frames with bees sitting on them (without a queen!) can come from different colonies. The youngest brood (eggs or one to two-day-old worker larvae) must be available on at least one brood frame, so that the queenless bees can make queen cells on them.

The breeding frames are hung in a box on the edge. Then a foundation next to it and at least a feeding frame. During transport, the three (or four or more) frames are secured with push pins against the last feeding frame.

The breeding frames are placed in a place that is outside the flight radius (ca. 3 km) of the bloodletted colonies c.q. the apiary. There, the foam rubber strip with which the entrance was closed during transport is opened a little on the side where the windows are hung. It is sufficient if the entrance is just one bee wide!

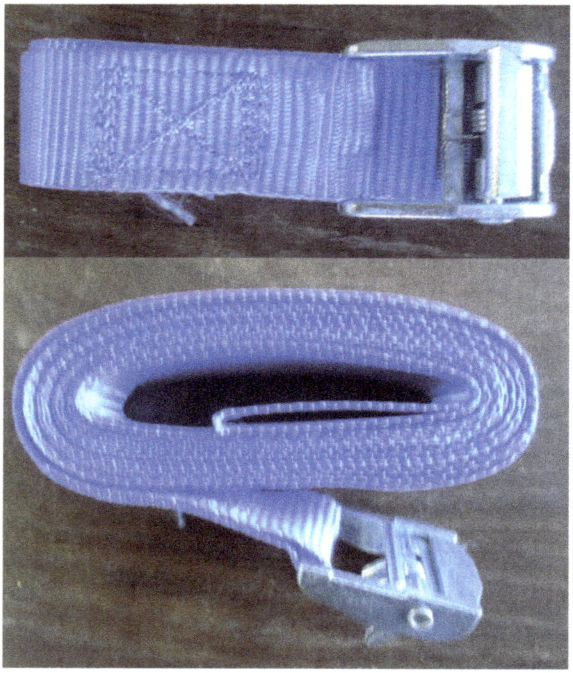

A very simple tension cord that can be obtained for little money and that is already very useful

If you do not have the 'luxury' of a second apiary about three km away from the apiary that provided the brood and the bees for the split you could use a (self made) "three frames box / nuc" for placing it outside the flying radius of your honey bees for about four weeks and bring this little box back to move the young colony in a hive box at your apiary. Family or friends living a few km away from your apiary usually will gladly house your nuc for about a month.

The split breeds a queen itself, provided youngest brood was available. If everything works out, the young queen will be laying eggs after about three-four weeks. The young colony can return to its place of origin if it is certain that the queen also lays the "correct" [= fertilised] eggs. One waits until the first brood is sealed.

A split with brood and bees can be occupied with an already fertilised queen nine days after its formation. The queen is placed in a plastic box with a feed dough closure in the split, after all the queen cells in it have been broken.

When adding an unfertilised queen, the risk remains that the queen will be lost on her mating flight. This risk, which is approximately 20%, also exists if one leaves it to the split to breed a queen.

When making a split with brood and bees, an empty frame covered with wire can also be hung between the brood frame and the feeding frame. Natural construction is produced there, as a rule, which initially consists only of worker brood cells. The corners of the frame are usually expanded later and then fitted with drone cells. This "mixed construction" can be avoided if, from the outset, one provides space for the split or the young colony with foundations.

The method of forming a simple split with brood and bees.

1. Prepare and deposit behind the colony:
    - a complete bee hive (mesh bottom box without varroa control inspection board and with closed entrance (foam rubber strip!),
    - brood box with a feeding frame and two foundations (one for the split, one to place in the brood box),
    - inner lid,
    - roof (lid),
    - push pins,
    - tension cord.
2. Turn on the smoker.
3. Open the colony, remove the honey box and place it to the side, remove the queen excluder and then:
4. Perform a "tilt check": tilt the upper brood box and check the combs for the presence of sealed brood and the expansion/size of the brood nest. Which frames are suitable for bloodletting as "breeding frames"? How well has the building frame been developed? Has the drone brood been sealed already?
5. Replace the top brood box, remove an edge frame and set it aside. Then loosen and lift the selected "brooding frame" with the hive tool. In addition, check the neighbouring frames for their suitability for bloodletting. Carefully check the "bloodletting frame" for the absence of the queen and make sure that there is youngest brood present that is suitable for making queen cells (eggs and one-two day old larvae in worker brood cells). This frame is hung with the bees sitting on it (i.e. without queen!) in the hive prepared as a split with brood and bees (on one side).
6. Carefully lift the drone brood frame (the building frame) (sliding the other frames next to it all at the same time) and also carefully check for the absence of the queen. Then shake off or wipe off the bees sitting on it in the brood chamber of the box as the split with brood and bees. Then replace the drone brood frame or install a new building frame.
7. Press all brood frames simultaneously against the building frame or the drone brood frame and place a foundation at the location of the removed brood frame.
8. Put the queen excluder back, put the honey box back on top and close the colony.

9. Place the foundation in the split next to the brooding frame hung on the side and then (at location three) the feeding frame. Attach the frames with two push pins (long drawing pins).
10. Close the split with foil, inner lid and roof, tighten it with the tension cord and place it outside the flight radius of the apiary.

There the entrance is opened again one bee wide.

(If the bees are less comb-resistant or if a three-frame box is used, it is wise to carry out steps 8. and 9. after step 4.)

Method of bloodletting with the formation of a collective split with brood and bees

1. Prepare and place behind the colony: a complete bee hive (mesh bottom box without varroa control inspection board and with closed entrance (foam rubber strip!), brood box(es), foil, inner lid, roof (lid), tension cord, a feeding frame hanging on the edge of the empty brood box, a sufficiently large number of foundations, a container with building frames for the exchange with sealed drone brood frames or a lockable container for the drone brood that needs to be cut out. Turn on the smoker.
2. Open the first colony, remove the honey box and place it to the side, remove the queen excluder and then:
3. Perform a "tilt check": tilt the upper brood box and check the combs for the presence of sealed brood and the expansion/size of the brood nest. Which frames are suitable for bloodletting as "breeding frames"? How well has the building framework been developed? Has the drone brood been sealed yet?
4. Prepare and place behind the colony: a complete bee hive (mesh bottom box without varroa control inspection board and with closed entrance (foam rubber strip!), brood box(es), foil, inner lid, roof (lid), tension cord, a feeding frame hanging on the edge of the empty brood box, a sufficiently large number of foundations, a container with building frames for the exchange with sealed drone brood frames or a lockable container for the drone brood that needs to be cut out. Turn on the smoker.
5. Open the first colony, remove the honey box and place it to the side, remove the queen excluder and then:

6. Perform a "tilt check": tilt the upper brood box and check the combs for the presence of sealed brood and the expansion/size of the brood nest. Which frames are suitable for bloodletting as "breeding frames"? How well has the building framework been developed? Has the drone brood been sealed yet?
7. Replace the top brood box, remove an edge frame and set it aside. Then loosen and lift the selected "brooding frame" with the hive tool. In addition, check the neighbouring frames for their suitability for bloodletting. Carefully check the "bloodletting frame" for the absence of the queen and make sure that there is youngest brood present that is suitable for making queen cells (eggs and one-two day old larvae in worker brood cells). This frame is hung with the honey bees sitting on it (i.e. without queen!) in the hive/box prepared as a collective split with brood and bees.
8. Carefully lift the drone brood frame (the building frame) (sliding the other frames next to it all at the same time) and also carefully check for the absence of the queen. Then shake off or wipe off the bees sitting on it in the brood box of the collective split with brood.and bees Then replace the drone brood frame or building frame (after cutting out the drone brood that has been cleared of bees).
9. Press all brood frames simultaneously against the building frame or the drone brood frame) and place a foundation at the location of the removed brood frame.
10. Put the queen excluder back, put the honey box back on top and close the colony.
11. Do the same to the other colonies as to the first colony.
12. When the container of the collective split with brood and bees is completely filled, close it with foil, close the inner lid and roof, tighten it with the tension cord and place it outside the flight radius of the apiary. There, open the entrance 3-4 fingers wide.

These interventions can be varied depending on the situation. If the colonies are first provided with a brood frame, the removed brood frame can be replaced by a second brood frame, which is then hung on the (other) side of the brood nest (place two or nine).

If no sealed drone brood is removed, the brood frame or frames can remain untouched and retain their worker bees. The number of bees sitting on the frame during the bloodletting/shrinking is more than sufficient for the care and keeping the brood warm.

If the beekeeper sees the queen, he has a free hand and can then work a little faster. He can temporarily put her in a cage or put her in a comb seam far away from the work place in the hive. In both cases, the queen must be caught by hand. This is possible and sufficient by grabbing her by the wings. On that occasion, she can also be marked if necessary. Those who are reluctant to catch the queen by hand and put her in a cage can also place the frame with the queen in front of the entrance and hang it back in the colony later, after having ascertained where the queen is on this frame, so that she is not in danger when the frame is hung back.

To get experience in catching the queen by hand one can practise with catching drones.

Drone brood cutting

The removal and excision of drone or worker brood cannot be described as a biological control method, and especially not if the brood is destroyed.

Removing sealed drone brood as a "varroa trap" is therefore worthwhile because drone brood is more attacked by mites than worker brood and the mites multiply much more strongly in drone brood.

The sealing time of the drone brood takes 2 to 3 days longer than that of the worker brood. Therefore, the mites in the drone brood generate more offspring. If many mites are given the opportunity to do so for a longer period of time, the mite number will increase rapidly. While with exclusive propagation in worker brood one can assume a monthly doubling of the number of mites, with propagation in drone brood every month a tripling or even a quadrupling is possible; of course as long as the bee colony is raising drone brood, which is usually the case from April to July.

The stronger contamination of drone brood cannot be traced back to an active preference of the varroa mites towards worker brood. The drone brood does not secrete any substance that is particularly attractive to the mites.

The reasons lie elsewhere. Two are known:

- The time period within which female mite females (can) penetrate the sealing-ready brood is twice as long in drone brood as in worker brood.
- Drone brood is visited and fed by nurse bees significantly more often in the larval stage than worker brood

Besides that it seems that varroa in Apis cerana, the Asian honey bee, from where the mites originate is anyway mainly found in the drone brood of that bee.

However, the apparent preference only becomes particularly visible when the drone brood is direct available for selection next to or simultaneously with the worker brood and both are in the appropriate stage (shortly before sealing). If a mite that is forced to lay eggs in a cell does not find drone brood, it will automatically go into a worker brood cell. It should also be borne in mind that in a normal strong colony there are many more worker brood cells than drone brood cells; for every two brooded building frames there are usually ten to fifteen frames with worker brood cells.

If the trap functions properly, about 50% of the mites are hidden in the sealed brood. This means (when multiplying the other mites in the worker brood) that with consistent cutting of drone brood, the mite number remains approximately constant during the "treatment". But even a smaller proportion of captured drone mites would be reason enough to cut the drone brood before the drones and the mother- and daughter mites that reproduce with them emerge.

The repeated excision of drone brood also has another advantage, so that the use of the "building frame" is worthwhile even if one does not think about varroa mites. The building frame helps to satisfy the "building urge" and the "breeding care" of the bees, which counteracts the onset of "swarming mood". It is an excellent colony mood barometer.

Drone brood

A higher effectiveness of the use is achieved if at least one "ready to catch" frame with open drone brood is constantly present in the colony.

The procedure of drone brood cutting can of course only be applied from April to June/July. After that, the bee colony usually no longer raises drones or only a few.

The recycling of drone cell construction (drone cell built)

The building frame wax can be used as additional wax recovery.
The building frame can also be provided with frame wire several times lengthwise and transversely to provide support for the construction of the cells. The brood is then unsealed with a knife and knocked out or washed out together with the mites. The advantage is that the colony can immediately be given an expanded drone frame that is immediately supplied with eggs. Naturally, wax production is then abandoned.
The washed drone brood can be used to feed fish or chickens. Chickens will then become your friends forever. When washing the drone brood in a suitable double sieve (which you need and use when extracting honey), the degree of varroa contamination can also be seen.

The care of the colony in spring and summer in short

What do I do?	What don't I do?
▸ during willow flowering check food supplies, if necessary provide feeding frames or additional feeding ▸ provide expansion boxes with six foundations and four clear extended frames ▸ increase size "from one to two" during willow flowering ▸ increase size "from two to three" during cherry flowering and provide the first building frame and place a queen excluder above the brood space ▸ provide the second building frame after about a week if necessary ▸ during good flow increase size "from three to four" around the flowering of the rapeseed	▸ do not change the mesh bottom box ▸ do not scratch open sealed food ▸ do not float feed ▸ do not provide additional feeding with protein dough ▸ do not rotate or exchange the brood

The "population increase in four steps" with integrated queen breeding

Modern beekeeping includes the annual breeding of queens. It can be effortlessly integrated with the population increase.

The new queens are used to provide the splits, of which you make about half as many each year as you have old colonies on the stand, if you want to keep the number of colonies stable. It is also useful to let young queens run into the old colonies in the autumn, because they will develop better the following spring with a young queen than with an old one. Until the change with young queens, these young queens, which are easiest to breed in May and June, but also in July, must be kept. Usually they are then kept in so-called fertilisation colonies in which the queens are already housed before the mating flight.

There is also no need to purchase expensive equipment for melting wax combs.

A metal sink tray that fits in a wooden frame with the perimeter dimensions of a brood box and a wallpaper steamer will work fine.

Attach four legs to that frame so that a bucket with some water can fit underneath.

Place a brood box on top with ten black combs, close it and fill the steamer with water. Start the steamer. The wax drips into the bucket. What remains in the frames after a while is the honeycomb contamination that can be scraped away. For further cleaning of the frames with hot water and soda, a second-hand dishwasher can provide good services.

Another disadvantage is that the egg production of the queen cannot be assessed after the mating flight. It is absolutely impossible to test the brood pattern. On the small frames, the queen finds only a few cells to lay eggs in. That is an unnatural situation for a fertilised young queen.

Most of the few cells in the fertilisation boxes are used as storage cells. As long as there is feed dough in the feeding compartment, it is taken and stored. Even if the combs are well supplied with food, the supplies can only be consumed for a limited time. The colonies can easily starve. That is why fresh dough must be constantly put in the feeding compartment. If the queen is removed from the fertilisation box, the remaining colony must disappear, which also means extra work. For these reasons, it is more efficient to keep all the fertilisation colonies for fertilisation on the stand in a box or in a container with Zander frames. This method avoids the disadvantages associated with using the smaller fertilisation boxes.

The population multiplication with integrated queen breeding in short

Day x	Form a collection brood box with nine brood frames with the honey bees that are on them and also sweep the bees that come from the drone frames if there are any and are cut out
Day x+9	Break all queen cells, make the collection brood box larger by placing a box underneath it. A breeding frame provided with breeding material [larvae of one-two days old] from a colony that qualifies for this because of its good properties (gentleness first of all) and then hang this breeding frame in the middle of the top box
Between **Day x+14** and **Day x+19**	Caging the sealed queen cells. With early caging on day x+14 (immediately after sealing the queen cells) it is possible to provide food in, preferably, a feeding frame tray. Caging must have taken place no later than Day x+19.
Day x+21 or **Day x+22**	Divide the brood colony by using a "bee frame" together with a queen each time to form the fertilisation colonies in boxes that are divided into four by three sliding walls and that are placed on one four-part mesh bottom tray each. Depending on the size of the food supply on the "bee frame" and the nectar flow situation, either a feeding frame, an empty built-out frame or a foundation is hung next to it.
Day x+36	Check whether the young queen has started laying eggs.

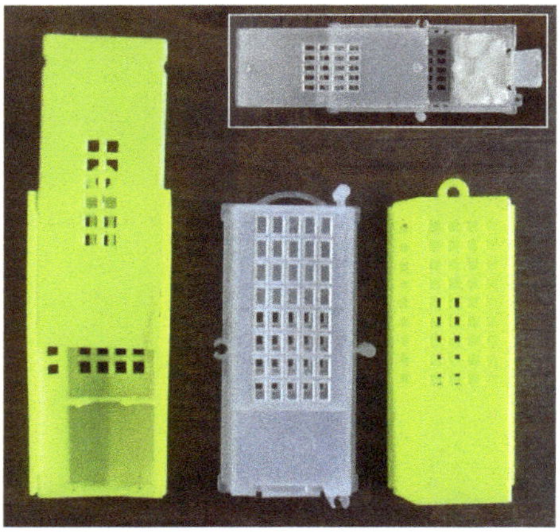

Plastic cages for queens.
If the cage is used to introduce a queen into a colony, the cage is closed with a wad of sugar dough.
(A piece of a marshmallow would do as well, for that matter).
Then it is also wise to put together with the queen a few attendants (workers) in the cage.

How to catch the queen

Catching the queen with your hand is not difficult at all. Finding the queen, on the other hand is often not that easy. But with experience comes skill. Most of the time the queen can be found in the middle of the brood nest, but also she likes to walk around on the building frame for the drones where a lot of young honey bees take care and are busy with feeding.

When you see the queen you capture her by pushing her (her thorax) gently with your index finger to the comb. You will see that immediately all the other honey bees will disperse.

Than gently grasp the queen between your thumb and index finger at her thorax, the firmest part of her body or at her wings. Never grab the queen by her abdomen. Now you can either cage her or tag her. Usually the latter means you have to transfer her to your other hand so you can use the tag or stick a plastic tag. Let the mark dry for a while before you put her back in the hive. Besides marking you can also cut off a piece of a wing to delay swarming (at most a quarter, better less), but it is more meaningful to exercise disciplined weekly swarm control. Never cut the wing of a virgin queen of course. If you only cage her (temporarily) add a few attendants as well.

You can practice this 'technique' quite well with drones!

The "four-piece mesh bottom box"

To use a one-box hive as a hive for four mating colonies, two associated components are required:

- a low special mesh bottom box that is divided into four equally sized compartments with three slit strip slats with a slot - each with its own flight opening - and that is therefore called a four-part mesh bottom box, and
- three thin sliding walls (partition walls) that are hung like a frame into the slit of a bottom slat and thus in such a way that two completely closed compartments are created. No bee may enter from one compartment into the other, not even at the top.

With a normal breeding series, one can count on at least twenty queens. If they all have to end up in the hives just described, one needs at least five of these boxes with a total of 15 dividing walls. This number is necessary as basic equipment.

The sides (the slats) of the mesh bottom box must have the same thickness as the walls of the box that is to be placed on top of it and they are (glued and) screwed together in the corners. Presumably such four-part mesh bottom boxes are not commercially available and the beekeeper will therefore make them himself or have them made. At the end of this book you may find several construction manuals.

The same mesh (stainless steel mesh) is used as mesh as in the regular mesh bottom boxes.

For each compartment, a small flight hole is drilled in the right place in each of the sides of the four-part mesh bottom box, so that each of the four colonies that are housed there will later have its own entrance on ("its") side of the hive. The round entrances can be drilled so large that the entrance of a queen cage fits in, which was previously placed around the queen breeding cell to prevent the young queens from attacking and killing each other. In this way, such a cage with a newly emerged queen can be inserted into this flight hole to provide the colony with a queen. (The entrances can simply be drilled small as well, to the size of one by one, and the queen can be offered to the colony in a queen cage with a sugar dough closure, or -with perhaps a little more risk- even simply the young queen can be allowed to enter the colony immediately after it has been placed in the compartment and the combs have been hung up).

A four compartment mesh bottom box. Actually it is a regular mesh bottom box in which three slats with a slot are attached in such a way that four compartments are created in an ordinary box by means of three dividing walls that fit into the slots.
Note that in the above depicted mesh bottom box actually a 'wrong' size mesh is used that would prevent an accurate mite fall control in a regular mesh bottom box. However, in this situation no control of fallen mites will ever be done.
Make sure that the four compartments are separated in such a way that no bee can escape to a neighbouring compartment.

In each compartment there is a small entrance. No honey bee will fly back the wrong way, nor will the queen when she returns from her mating flight.

‹‹‹ In the figure you see such a four compartment box with two of the four entrances at each side visible.

The box with the four-part mesh bottom box is filled with three partitions and four times two frames. The foil placed on top prevents the bees from walking from one compartment to the other, but of course this only works if the foil cannot be pushed up by the bees. The inner lid must therefore lie completely flat on top.

Queen rearing in the queenless foster colony

The occupation of the box with four fertilisation colonies above a four-part mesh bottom box with inserted partition walls is particularly successful if the queens are reared in a queenless colony. In addition to the method described, there are many others. Most methods are so complicated that they rather discourage queen rearing. The method shown here is simple and can also be varied. Feel free to try it out. If it works, which depends not only on your beekeeping skills but also on the weather, you will have a lot of fun with your young queens and their young colonies. And if it doesn't work the first time, simply try again. A collective brood layer can serve as a foster colony, which, as already described, is obtained by bloodletting/shrinking the production colonies. One can also sacrifice a strong production colony. There are three (and more) possibilities for that alone.

The first is: one looks for the queen and kills her. The colony starts to make brood cells. After nine days the brood cells are broken and the breeding frames are hung. This method leads to the same result as a collective brood layer.

The second is: one looks for the queen and takes her away with a (brood) frame that is well occupied with bees. This "queen layer" is housed with one feeding frame and a foundation that is hung between these two frames in a box with two transport guards (two pin tacks), placed in another position (at least three km away from each other) and then cared for as a regular split. This split also grows in the summer into a colony that is strong enough to be able to overwinter.

In the queenless remaining colony, all the queen cells are broken after nine days and then the breeding frame is hung up. The second variant does require more work, but at least you start a new colony with the caught queen as well.

The third is: when removing the queen, all the frames with open brood are also removed. This also includes the frames that mainly have sealed brood but also some open brood! The bees are rejected or wiped off and the brood frames are placed with other colonies. (You can also make splits from them). When removing the brood, the colony is placed on two smaller boxes. In the middle of the upper box, space is left free for the breeding frames. One or two hours after removing the queen, the breeding frame with larvae is hung up. This can also be done one or more days later, but then a frame with open brood must be hung in the foster colony in the meantime so that it has the opportunity to make queen cells there.

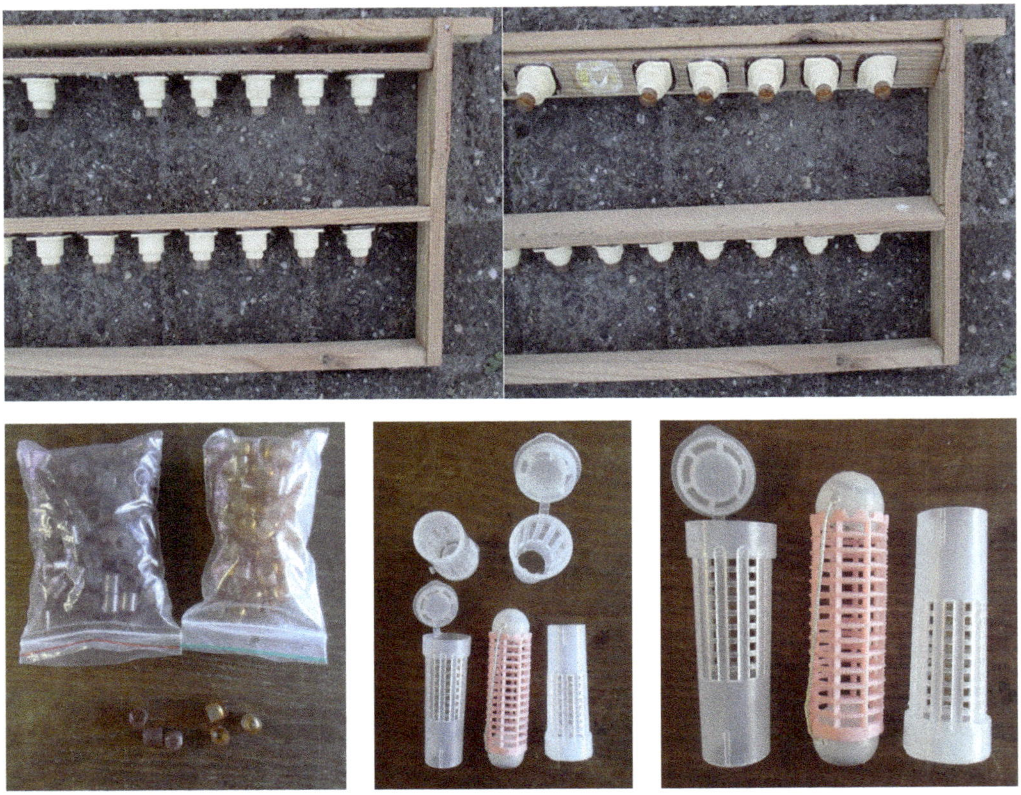

So, a breeding frame is needed to put in the foster colony that you have created. The plastic cup holders are glued to the preferably rotating crossbars. A very young larva is placed in each cup which is first placed in the holder. In the past a cup or bowl was made by dipping a stick of the right size in liquid (hot) wax and than attaching it to the crossbars also with liquid wax. The modern plastic cups save you all that work.

When the queen cells are sealed you have to re-cage all the developed queen cells with these 'hair curlers' that fit on the cup holders, so that no queen will kill the other ladies. Often the young queen crawls back into her pupa cell and suddenly doesn't know how to get out again. The remedy is to put already a few attendants in the cage also that can and will free her and keep her well.

A fourth variant is: the already queenless colony keeps all its brood and the breeding frame is nevertheless hung up one to two hours later after the queen has been removed. This entails the risk that the bees will predominantly prefer to use their own open brood to make queen cells rather than the foreign brood that the beekeeper has given them with the breeding frame. The acceptance percentage of larvae in the brood cups can be between 0 and 80% as the highest value that is often achieved when breeding in queenless colonies that no longer have their own brood. From the queen cells made on the brood frames, young queens can emerge after only nine days if the larvae chosen by the bees were older than the larvae that the beekeeper has placed in the brood cups of the breeding frame. Therefore, the queen cells on the own brood must be broken in time.

Hanging breeding frames immediately after removing the queen from a production colony (variants three and four) or immediately after forming a collective brood layer (variant five) has another serious disadvantage: when the foster colony is taken apart twelve days after its formation when the young queens have emerged, it still has sealed brood. In that case, a varroa control treatment in a brood-free state is not possible.

Sacrificing a production colony to form a foster colony is somewhat 'problematic'. That sacrifice costs one or two honey harvests. So, preferable is mostly to take the route of the collective split with brood and honey bees that is hung in the breeding frame nine days after its formation or receives it after the queen cells have broken. Its age structure ensures the success of queen breeding and offers good starting conditions for the mating colonies, as they can be successfully treated against the varroa mite when they are created in a brood-free condition.

Method of hanging in the breeding frame on day x+9

- Lift the collection brood box from the mesh bottom box and place it on an inverted metal roof or on a second mesh bottom box,
- Place a box for making the bees bigger on the mesh bottom box of the collection brood box (with nine half-full feed frames or filled feed frames and empty frames),
- Lift a (brood) frame from the brood box, shake the bees onto or into the box for making them bigger, carefully check the frame for queen cells that must then be destroyed, then hang the frame in the box for making them bigger,
- Place an empty box on the box for making them bigger,
- Lift the remaining frames from the collection brood box and, after shaking off the bees and breaking the queen cells, hang them in the "empty box" placed on top,
- Leave space in the middle of the "empty box" for the breeding frame(s) between two frames with sealed brood,

- Cover the (so formed) foster colony with foil, inner lid and the metal roof until the breeding frame or frames with larvae are hung in place.

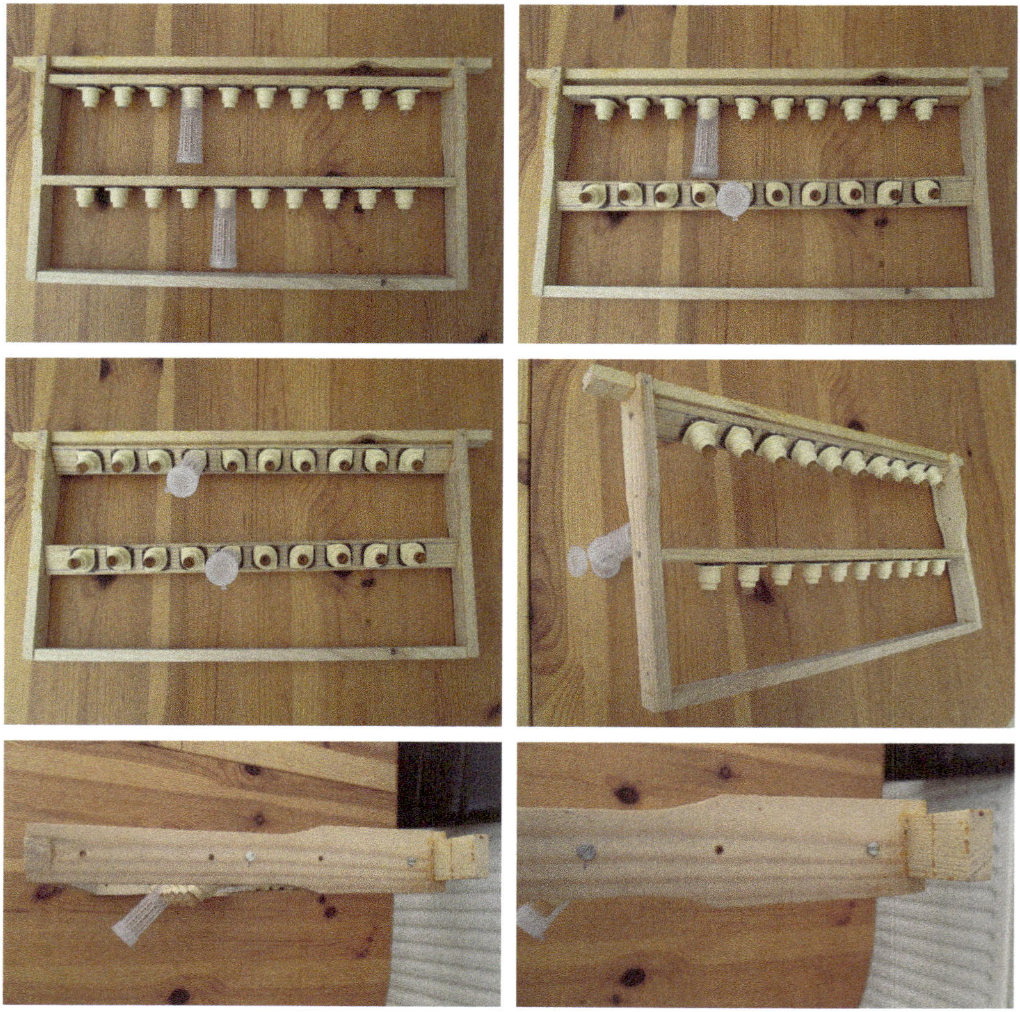

The breeding frame with the cup holders glued on the cross bars. The cross bars can be turned to make it easier to put on the cages when the queen cells are capped. To obtain this movable possibility the cross bars are attached to the sides with just one screw.

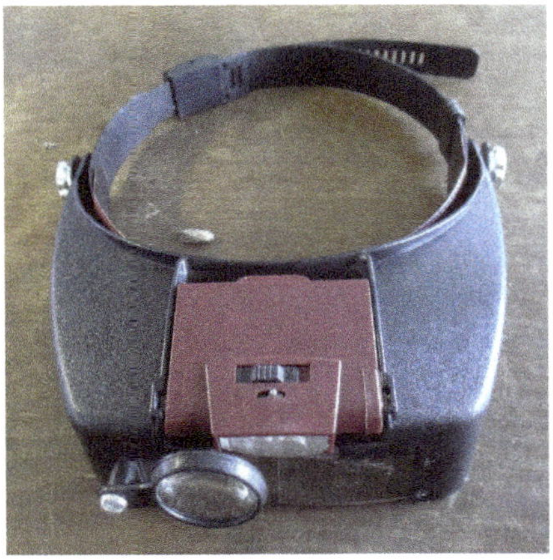

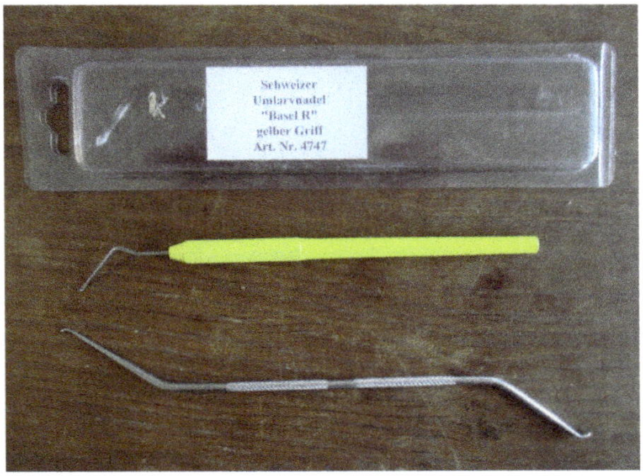

Besides a larval needle you can simply use a 3+ (or higher) pair of glasses. For a better light in the cells and with some + or - adjustments to your eyes and various magnification options a head loupe is not a luxury. To get the larvae out of their cells into the (plastic) cups you need a larval needle. These come in different designs, but the Swiss larval needle (the one in the picture with the yellow handle) is a favourite.

The breeding frame

A larval needle and a breeding frame are needed for queen breeding. In addition, run-out cages (the size of hair curlers) are needed for the cages of the queen cells.

The breeding frame is an empty frame in which two (rotatable) cross-bars (breeding slats) have been made. The special holders for the breeding cups are glued onto these breeding slats. Up to twenty cups can be placed on a Zander format breeding frame. Perhaps however it is best to start with sixteen or even thirteen cups so that you are sure that you will have enough space to place the run-out cages.

Suitable are plastic cultivation cups-pans that are commercially available as complete systems (with holders and run-out cages).

You can also make your own cultivation cups from wax; melt the wax and dip a stick with a round end of the right size into the melted wax, remove the stick and knock off the cultivation cup thus formed - if necessary, briefly hold the stick with the wax in cold water. However, this time consuming procedure does not outweigh the little costs of the cups that are commercially available.

From which colony should you obtain breeding material?

As a breeding material supplier, a production colony is selected that has not been negative up to that point. The most important criterion is gentleness. High standards must be set for this. Do always breed from very gentle colonies.

However, those colonies that have suffered from chalk brood must be excluded from the gentle colonies. The one-time occurrence of chalk brood mummies completely disqualifies a colony. Other criteria such as "no or little desire to swarm with good honey yield" only come in third place when choosing the "breeding queens".

The result of this selection cannot be predicted in advance; because the characteristics of the young colonies are not only determined by their queens, but of course also by the heritage of the drones that are involved in the fertilisation of the queens. Which drones these are cannot be predicted in a natural apiary fertilisation.

For the larvalisation, a brood frame is taken from the selected colony on which there are many youngest, just hatched worker brood. These may not be older than one-two days. Only then can they be turned back flawlessly into queen larvae by the nurse bees of the foster colony.

For the larvalisation, good light, a steady hand and a good eye are necessary. The larvae are less vulnerable on the back than on the belly. Therefore, the larval needle [larval spoon] - on the picture it is held in the right hand - is carefully placed behind the larva on the cell bottom (seen from the viewer's point of view, in reality it is the back wall of the cell) and then guided under the larva so that it comes to lie on the needle (spoon) with some food juice. When placing the larva in the breeding cup – which in the picture is held in the left hand – one also proceeds as follows: place the needle on the bottom and pull it backwards so that the larva is scraped off. When all breeding cups are provided with larvae, these are put in the cup holders. The breeding frame is then hung in the middle of the upper box of the foster colony and the box is closed.

The open brood is sensitive to UV radiation. The breeding material frame and the breeding cups with larvae may not be exposed to the sun during the larvalisation. Use a slightly wet towel to cover the brood frame as much as possible.

The queen cells are caged at the earliest after 5 days and at the latest after 10 days after the larvae have been transferred ("provided with walk-out cages") to prevent the queen that appears first from killing her collaegues. If a plug system is used for queen rearing, the walk-out cages (filled with 3-5 accompanying bees) can simply be slid over the queen cells, so that they remain on the breeding bars. Cells whose holders come loose when the breeding bars are caged are placed above a comb alley of the foster colony.

Gerhard Liebig in larvalisation action, also making use of a head loupe. The head loupe is also useful for the diagnosis of the mite fall, so that all fallen mites can be recognised.

Photo: Dr. G. Liebig

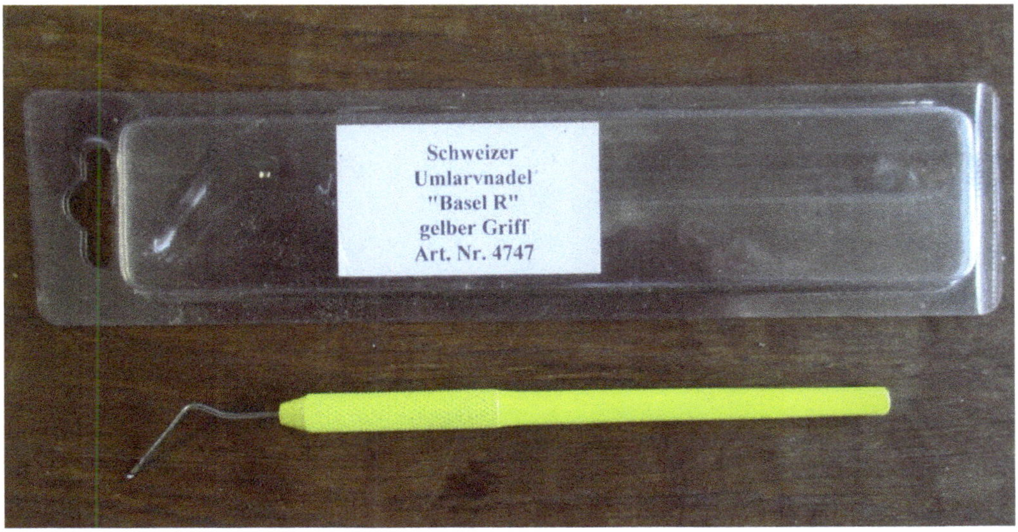

Caging the queen cells on day x + 19

With a strong foster colony, one can count on about 75 % queens being reared out of al the larvae offered.

A breeding frame for the cage of the queen cells. Both breeding laths are provided with holders and breeding cups made of plastic. In some cups - like above in 4th and 5th from the left - the larvae were removed.

Photo: Dr. G. Liebig

The same breeding frame as in the previous photo - of course seen from the "wrong" side (so turned verticall)] - after caging (with run-out cages). Four of the 30 larvae were removed.

Photo: Dr. G. Liebig

The queen cells are caged after five days at the earliest and after ten days at the latest to prevent the first-born queen from killing the other queens. Cells whose holders come loose when the breeding frame is raised are placed above a seam of the foster colony.

In case of good nectar flow, the breeding frame is filled with wild growth after sealing the cells. Before caging, the built-in queen cells must be cut free with a sharp knife heated above a flame (cigarette lighter, gas burner or whatever else you use to light your smoker) without damaging them.

When breaking the queen cells and immediately before hanging the breeding frame, the foster colony, if it has emerged from nine brood frames, is given a box with built-out (feed) frames underneath, so that twelve days later, when the worker brood and also the queens have emerged from the previously "caged" queen cells, the honey bees of the foster colony occupy nineteen frames and the breeding frame. Then, with about fifteen emerging queens, fifteen fertilisation colonies can easily be made

The queens that have emerged are brought into the fertilisation colonies as the queen. But first the breeding frame is lifted and the number of queens that have emerged is determined. The cells where no queen has emerged yet are taken out of the run-out cage and stripped of the holder to check whether the queen that has not yet emerged is alive or has died. To this end, the queen cell is carefully opened on one side near the cell bottom with a knife and opened. If the queen is alive, the cell is pressed back into the holder and given as a cell with a queen that has not yet emerged in a fertilisation colony. If not, it is rejected.

The decision to remove queen cells from which no queen has emerged is made when more queens have emerged than fertilisation colonies can be formed. The number of possible fertilisation colonies is assumed by the number of "bee frames". A foster colony on one box provides the bee material for ten fertilisation colonies (= ten bee frames - if, in addition to the fertilisation colonies with the nine frames, the breeding frame of the foster colony on one box can also be used for the delivery of bees). With a foster colony on two boxes, twice (or two x nine + one) as many fertilisation colonies can be formed, but then sufficient queens must have emerged.

Making the fertilisation colonies

The fertilisation colonies are quartered, each in four, in the compartments of a box with a four-part mesh bottom box or, each for itself, in a box with a mesh bottom box. The boxes are prepared. Their number depends of course on the number of caged queen cells.

With closed entrances, a frame of the foster colony occupied by bees is hung in each compartment of the four-part mesh bottom box box. Care is taken to ensure that the "honey bee frames" have approximately the same number of bees. Usually, the central frames in the bottom box of the foster colony are denser and occupied by more bees than the edge frames or the frames in the top box, which nevertheless contain more food. If it is necessary to distribute the bees equally, each bee frame must contain at least a thousand bees (this is the case if one side of the frame is completely occupied with bees (see also under 'population estimation'; there are more if these bees or some of them stick their heads in the feeding cells), then first the central frames are lifted out of the lower box and the surplus of bees are shaken off onto the frames that are less occupied (edge frames in the lower box or the frames in the upper box). Sometimes it is also sufficient to wipe off the cluster of bees hanging underneath.

Before hanging the "bee frame" in the compartment of a four-part mesh bottom box (or in a box), the colonies can be treated with oxalic acid against the varroa mite by spraying the combs occupied by bees on both sides, which is not necessary if the foster colonies are only mildly infected. A fall diagnosis carried out in advance serves as a decision aid. An alternative to the spray treatment of the "bee frames" when creating fertilisation colonies would be the drip treatment of the foster colony one day before it is taken apart.

A well-filled honey or feed frame and the queen (or sealed queen cell) are added to the "bee frame". If the "bee frame" contains a lot of food, an empty frame or a foundation in the compartment of the four-part mesh bottom box can also be given as a second frame.

The inner lid is placed on the foil, with its flat side facing downwards, so that it fits well on all sides. This prevents the bees from building a superstructure on the frames, pushing the foil upwards and thus gaining access to the neighbouring colony. The metal roof is placed on top of the inner lid.

When placing the young queen, there is a risk that she will fly up and away. Caution is therefore advisable. It is best to place or stick the opened cage on or in a comb seam of the fertilisation colony and then cover it (immediately) with foil until the queen has left the run-out cage and has dived down into the comb seam. This can be done simultaneously with the four fertilisation colonies of a four-part mesh-bottom box. A variant would be to pull the cup with the built-out queen cell out of the run-out cage and stick the cage thus opened in the entrance of the relevant compartment of the four-part mesh-bottom box. Once the queen has left the cage, it is pulled out again and the entrance is closed (again) with a piece of foam.

From a foster colony on two boxes, five four compatment hives with a total of twenty fertilisation colonies can be filled by simply hanging the frames (actually there are only nineteen frames, but the bees from the breeding frame are thrown off in a prepared compartment with two frames).

A special treatment of the bees as when filling fertilisation boxes is not necessary. The presence of drones is tolerantly accepted, since they are of no significance for the fertilisation result of the queens in these fertilisation colonies.

The hives are taken that same evening to a fertilisation location outside the flight radius of the foster colony up to now. A cellar lock-up is not necessary. During transport, make sure that the colonies are provided with sufficient ventilation from below. This is guaranteed by placing the mesh bottom boxes on wooden slats.

At the place of fertilisation it is advisable to place the boxes on relatively higher coasters (wide planks, stone slabs) so that there is something like a "landing place" in front of the entrances and the entrances do not grow closed with the grass or weed around. For the necessary control it is also better if the boxes are a little higher and not on the ground. In addition, one must maintain sufficient distance between the boxes to prevent the queens from flying away or to the 'wrong' hive.

As soon as the queens are fertilised, there are sufficient empty cells available to them on the two frames to lay eggs in, so that the brood nest grows quickly and at the earliest two weeks later it can be assessed whether it is closed or has shot brood. With this control it is also little work to test the size, shape, colour and integrity of the queen. On the two frames, sometimes she also stays on the box wall or the sliding frame, she is quickly found and can then also be caught and marked.

As mentioned earlier: catching and marking the queen(s) can be practiced with the drones.

The easiest way then is to grab them by the wings and hold them between the fingers of the hand you don't need for marking. With a marker a colour mark is then dotted on her back. Then she can be put back in the colony or put in a cage for a while to let the paint dry.

Not all queens start laying eggs, because some of them do not return from their mating flight, for example. Queenless colonies are united with the neighbouring colony by pulling out the sliding wall. The queen is first secured in a cage with food dough as a barrier so that she does not suffer any damage during the time that the bees of both colonies need to get used to each other and in which they eat the queen free. The frames of both colonies are hung together. If there are already drone mothers (egg-laying workers) present, the colony in question is swept or shaken off a few metres away and the sliding wall is lifted. In this case too, the queen of the neighbouring colony must be protected and caged so that she is not killed by any drone mothers that may return home. (The further away from the hive location one can push off or sweep off the frames, the less chance there is that the drone mothers will return to the hive).

There is no need to lift a frame. The superstructure under the foil discloses it: all four fertilisation colonies are in order and must be taken out quickly.

Photo: Dr. G. Liebig

As soon as the first brood appears, the colonies start to grow. Then it can quickly become too cramped on two frames. That is why the colonies must be moved in time to a box with a mesh bottom box. This transfer is connected to a move (of location). In new colonies building that have started early May or early June transfer is absolutely necessary if the growth potential of the young colonies is to be fully utilised. Until the fall, the colonies fill at least one box. If these young colonies are continuously fed, they build eight foundations as full combs until autumn. In the beginning, they can also be given empty frames with only wire covering next to or in the brood nest. As long as the colonies are small, they fill these frames exclusively with worker brood construction. Only later, when they occupy more than four frames, drone construction is also made. With later queen breedings that started in July, it is no longer to be expected that young colonies will emerge from fertilisation colonies that can overwinter. Their queens serve primarily to replace old queens from the production colonies in the autumn. When this is the case, three of the four queens are removed and all the partition walls are lifted to unite her bees and brood. The remaining queen must be caged in advance! Bees and brood from four colonies usually result in a young colony that can overwinter successfully. The hive is filled with two feeding frames, the colony with closed entrances on a another place outside the flight radius where one entrance is opened again.

On occasion the colony is placed on a normal mesh bottom box and of course also fed. The fertilisation colonies of early built can be housed in a hive immediately. This saves you the quartering but you have more work with failure.

Dismantling the foster colony on day x+21 step by step

- Prepare as many hives with mesh bottom trays or four-part mesh bottom trays and have the built-out (feeding) frames ready as queen cells were placed in run-out cages two (-seven) days beforehand. If fertilisation colonies are housed individually in a normal hive (with a mesh bottom box), then each also requires a foundation and transport locks (pin tacks). The flight holes are closed with foam. Tension cords are also required for transport.
- Place the hives around the foster colony.
- Light the smoker, open the foster colony, remove the breeding frame, assess the bee population on the neighboring frames, loosen all caged queen cells and after checking whether or not the queen has emerged, place them sorted on an inner lid.

- Open cages with cells from which no queen has emerged, remove cup with queen cell and check its condition. If the observation is "queen healthy or shortly before emerging" then close the cell again and put it in the run-out cage.
- Prepare acid treatment of the frames occupied by bees: put on gloves, fill the sprayer with prepared oxalic acid solution, put on goggles and put on the mouth guard.
- Remove the upper box of the foster colony and place it on an additional mesh bottom box.
- Remove a well-occupied frame from the lower box of the foster colony, assess the bee occupation, if necessary slide the cluster of bees hanging below it off above the upper box or shake off or wipe off the obvious "excess" of bees above the upper box.
- Spray the bee frame on both sides with the acid solution and hang it in a compartment of the first four-part mesh bottom box hive (or in a hive), fill the compartment with a suitable second frame (with or without food), place a run-out cage with a young queen on the comb alley of the compartment.
- Proceed accordingly with the next three frames of the foster colony, then the run-out cages are opened one after the other and inserted into a entrance of a compartment, if it has a size that matches, so that the queen can enter the compartment and the entrance is closed again. If the run-out cages cannot be inserted into a entrance, the young queens are allowed to enter a comb alley of her fertilisation colony; it is important that they do not fly up and away when leaving the run-out cage.
- The four-part mesh bottom box hive is closed with foil and inner lid and prepared for transport to the new location by applying a tension cord. Then the next fertilisation colonies are quartered in "four-part mesh bottom box hives".
- After dividing all the bee frames, spray the bees that are on the hive walls with an oxalic acid solution, then push these bees into the bottom box, hang a feeding frame, a built-out frame and a foundation on the edge, make the entrance there smaller, add a young queen or brood-ready cell, close off the colony.
- Bring the fertilisation colonies to their location.

Brooded natural comb construction of a young colony. One saves on foundations, but often "mixed construction" of worker (brood) construction and drone (brood) construction occurs.

Photo: Dr. G. Liebig

Caring for young colonies

Young colonies can only grow if they have sufficient food at all times. They provide their own pollen supply. The nectar supply is often so low in the summer that the young colonies consume more than they can find in nature. As a rule, the young colonies must therefore be fed continuously. Apart from feeding and timely addition of foundations, the young colonies do not require any further care in the summer. In late summer (August/September), they are raised and treated against varroa, just like the production colonies.

The entrance of splits and young colonies must be kept small in principle. Feeding is only done when the queen has started laying eggs. This does not have to be checked continuously, but (as a rule) only once: take into account the development time until sexual maturity, wait for the right weather (ideal: sunny, calm, 20°C) for the mating flight and give the queen time to have provided a large number of cells with eggs. It usually takes about thirty days "from egg to egg". If you wait five weeks, it is sufficient to lift a frame. If there is brood available, feed with sugar water or with feed dough. The feed dough can be placed as a flat sheet on top of the frames and covered with foil and the concave side of the inner lid. After two to three weeks the young colony is checked and feed dough is placed down again.

Sometimes it is necessary to check if there is still an (egg laying) queen in the colony. Then a frame with also youngest brood (without the queen...) from another colony can be placed in the colony to see whether queen cells (supersedure) are formed. If it is not the case (and also there are no drone eggs laid -pile brood-) than apparently everything is still fine in that colony.
Once you are good at larvalisation, it is also convenient to press a few cups attached (glued) to a paperclip with a youngest larva in it into a (brood) comb of a suspect colony instead of a frame with brood.

Of course, this is also an excellent way to create a new queen in a colony if it is necessary or desired to replace the old one.

Sugar water (or syrup) was earlier given to the growing young colony in Tetra packs that are placed on the mesh bottom against the occupied frames. By now the use of Tetra packs is abandoned, because there are always colonies that gnaw a hole in the bottom of the pack. A flat plastic dish -of vegetables or meat products from the supermarket- on the mesh bottom with syrup or sugar water and a handful of "vegetables"(= twigs, leaves or grass) -so that the bees do not drown- works better. So, don't forget the swimming aid in/on the sugar water! More elegant than using these plastic dishes are the closed feeding frames that are either one or two frames wide and can be filled with feeding dough, honey, syrup or sugar water for feeding. Don't forget also in these devices a swimming and climbing aid either!

The honey harvest

After the end of the blossom period in spring, the honeycombs can be removed; Obviously only when the honey is ripe and contains little water.

The water content is the most important quality characteristic: the drier the honey, the better its quality. Therefore, the beekeeper must correctly estimate the water content before removing the honeycombs; because if he only notices during or after extracting that the honey contains too much water, it is too late.

Extracted honey cannot be dried. Honey with a water content of 18% or more has a limited shelf life. It quickly goes into fermentation and may no longer be sold. Personal use of such honey is also discouraged. Such honey can only be given to the bees. When it comes to that, a lot of work has been done for nothing. Then it would have been more advisable and simpler not to extract the honey at all.

How can one tell whether the honey stored in the combs is ripe? There are some starting points or indications, but none of them provide absolute certainty. First of all, a distinction must be made between sealed and unsealed (open) honey.

The honey is ripe when it is sealed, as stated in many manuals or trade books. But that is often not the case at all. One cannot and should not rely on it.

The cells are only sealed by the bees when they are filled with honey. As a result, there would still be room for more honey in the unsealed cells. When the flow comes to an end and the colony produces less from nectar or honeydew, these cells remain open and are even emptied.

In the time after the end of the flow, the open honey dries. The longer the beekeeper waits before removing the combs, the drier the honey becomes in the open cells. However, this process continues as long as there is no flow. As soon as there is flow again and the bees bring in again watery nectar or honeydew, it is piled on top of the already dried honey, which increases its water content.

The honey in the sealed cells can also still dry, although the washing lid of the cell strongly prevents both drying and water absorption. If the colony has sealed a honey that is too moist, the subsequent drying in the colony takes considerably longer than the open honey that has not yet been sealed. Therefore, it may happen that at the time of honey harvest, the open honey contains less water than the sealed honey.

Harvest only ripe honey

The water content of the honey can differ from comb to comb.

The following rules apply:

- The honey is moisture in the combs at the edge than in the combs in the middle
- In the combs of a fourth box (second honey box) the honey is more moist than in the combs of the third box (the regular super or honey box)
- In thick, broadly developed combs the honey is more moist than in thin or normally developed combs

These rules must be observed when checking the water content. The water content is checked where it contains the most water in accordance with these rules. To test, a particularly thick comb is selected at the edge of the upper container. If the first test is positive (= "the honey is dry"), the result is checked with another comb. There are refractometers on the market that allow you to precisely determine the water content of one drop of honey. A practical alternative is using a coffee spoon, the sealed honey cells are scraped open on a particular comb and enough honey is collected to fill the spoon half way. The honey is then allowed to drip from the spoon and it is assessed how or how quickly the honey flows from the spoon. You can practice this at home during breakfast with the honey that is on the table and which you know has a low water content: just practice!

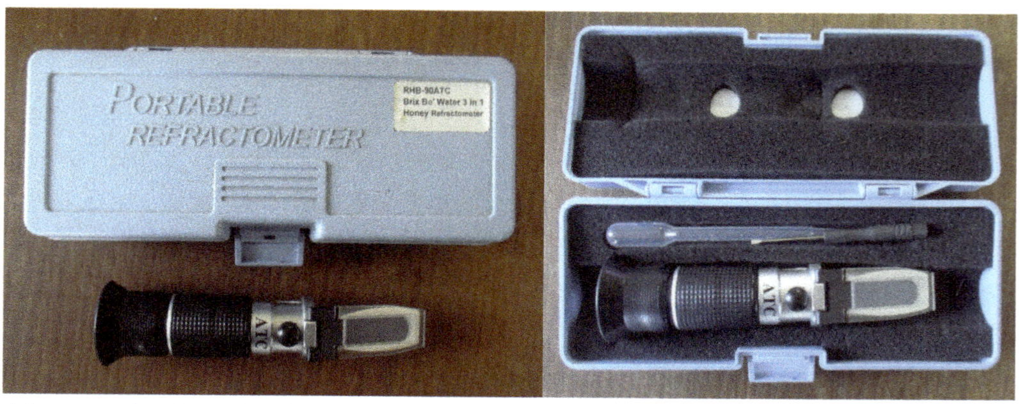

A honey refractometer. If you purchase a refractometer be sure that it is fit to determine the water content percentage of honey as there are also refractometers for syrups or liquids with sugar.

The temperature also influences the dripping or flowing behaviour of the honey. This must be taken into account when testing. The warmer the honey is, the less its consistency.

Honey freshly collected from the colony has a higher temperature than honey from a glass jar in the pantry and therefore flows faster.

A knife can also be used instead of a coffee spoon. The honey sample on a knife point should not fall off when the knife is continuously turned, but should allow itself to be wrapped.

If the honeycombs are not completely sealed, the water content of the open (!) honey can be checked with the "spray" or "impact test".

If drops of honey squirt from the comb, the honey is not ripe enough and the harvest must be postponed to a later date.

One drop is enough to make the decision!.

Extracting the unripe honey always reduces the quality of the honey. Therefore: let the honey mature in the colony!

Problems with the water content are more common with blossom blooms than with honeydew honey. That depends on the basic material. Nectar usually contains much more water than honeydew. The more moist the collected material, the longer it takes the honey bee colony to dry the honey.

The requirements for the harvest of ripe honey

Only harvest the honey

1. **after the end of a mass flow**, e.g. after the rapeseed flowering (in spring) or after the linden flowering (in the summer) after which only "babble flow" has prevailed;
2. **after a bad weather period** that led to a flow pause of several days during the " babble flow ";
3. after this multi-day flow break caused by the weather conditions, **early in the morning**, so that the honey that was introduced the day before has already thickened during the night and the small amount of nectar already collected immediately has not yet been stored in the honey box;
4. if the open honey does not squirt,
5. i.e. **not a drop falls out during the "spray" or "impact test"** with a comb with (much) open honey.

With brush or bee escape

If the honeycombs are removed after the end of a flow or during a flow break, robbery can easily break out. Then the bees can attach themselves to the combs faster than they can be swept off. The bees repeatedly "treated" by shedding or wiping become aggressive and eager to sting. The beekeeper can protect himself from them by wearing a thick disguise, but the honey harvest, as well as all other interventions in a colony, provide much more joy if smoke and a veil are within reach as "certainty". Robbery and stinging only result in sweating and swearing. This should not be the case, not even if the honey harvesting is done easily and correctly.

The honey harvesting is very simple if the honey box is separated from the brooding area by a queen excluder. There are then two options: either one removes the honeycombs one by one from the honey box and shakes or wipes off the bees that are sitting on them, or one places a bee escape instead of a queen excluder or places a bee escape on top of the queen excluder so that the bees leave the honey box forever. A renewed access will be denied.

After a night with a bee escape placed, the honey box is free of bees. Indeed, a little more time elapses between the introduction of a bee escape and the honey harvest if one adheres to the requirements for the harvest of ripe honey and therefore places the bee escape in the early morning and then removes the honey box in the evening of the next day.

When applying a bee escape, the apiary must be visited twice. With the classic honey harvest with the brush and "comb by comb", one visit is sufficient. Before the first colony is opened, a wheelbarrow with an empty hive box (container with closed bottom and lid) is placed behind the ready-to-harvest colony, in which the honeycombs that have been released from the bees are placed and a second box containing empty combs and foundations. These will replace the removed honey frames.

First, accompanied by a few puffs of smoke, the lid and foil are removed. Then frame by frame is lifted and the honey bees on it are swept off, but not in the colony, because then you will have to deal with them again, but in front of the entrance. Required is that you can work in the colonies from the side and not stand on a trestle. That is als a good reason to group always the colonies in pairs. If the colony is not already put on a broad plank that also serves as a "landing site", a foil or inner cover is placed in front of the entrance on which the honey bees are swept. It takes some time before they notice what has happened to them and return to the colony via the entrance. Then however the honey harvest is long over.

The full honeycombs in Zander- or Langstroth format weigh between 2 and 3 kg. This weight will eventually be noticed in your arm when the combs are held up in the hand by one ear of the frame and wiped with the other hand. This is quite a burden for the lifting arm. This is not noticeable with the first colony, but it is noticeable with the third or fifth colony. This burden is significantly less if the frame with one ear is placed on the frame carrier at the front of the box, while the bees that are swept away.

In the old days so-to-say, a goose feather was used to gently brush the honey bees from the comb. The feather depicted is actually from a crane, for that matter. Nowadays, a soft brush is often used, also because it is easier to clean, especially if there is honey on it. Wipe the bees from the comb in front of (the entrance) of the hive.

An alternative that does not require a feather or brush is to use a bee escape that fits in a round hole in a plate that is placed between the honey box and the broodbox (in stead of the queen excluder, but I prefer to put the queen excluder back under the bee escape as well). The honey bees that return from the honey box to the brood box cannot go back to the honey box. If the bee escape is used overnight, the the next day the honey box will be found without bees.

If a lot of space has already been freed up in the honey box, the last frames to be brushed can also be placed on the frame carrier at the back of the box and the honey bees can be swept onto the queen excluder. They then usually crawl downwards into the brood box and do not fly back to the honeycombs that are still in the almost empty honey box. If they do, they will need to be wiped a second time, which can lead to react with an increased desire to sting.

If the honey is to be harvested from two boxes, all bees must be swept from the top honey box in front of the entrance. One can also act differently when harvesting honey "with the brush": the full honey space is removed and placed behind the honey bee colony on the ground or raised on a low box tower that is covered with an inverted metal roof. The colony is given an empty box on top of the queen excluder in which the bees are wiped from the full honeycombs and which is filled during or after the honey harvest with empty combs, foundations and any (edge) frames that are not completely or only slightly filled from the just emptied honey box. When harvesting honey, care must be taken to ensure that the honey bees do not have access to the honeycombs that have been released from bees.

For that matter: the bees can be wiped off faster with a large brush than with a small one. If possible, it is perhaps more preferable to use the bee escape than the brush. A way to use a bee escape is to prepare an expansion box for each colony, which is equipped with a comb core and three foundations on the right and left. While paying attention to the requirements for harvesting ripe honey, the honey box is removed using also the smoker and set it aside. Then the expansion box is placed on top of the queen excluder on which the bee escape is placed, on which the honey box is then placed back (and the honey bee colony is covered with [foil, if possible] a lid and roof).

When taking away the bee-free honey boxes the next day, usually in the evening, no smoker is necessary. The bee escape can also be removed without using the smoker. The bees that are on the underside are repelled in front of the entrance. The bee escapes can also be placed in front of the entrances and then collected when all the colonies on the apiary have been harvested and the boxes with honeycombs have been put away in the car or locked away somewhere else. It only takes a few minutes for the honey bees to leave the bee escape placed in front of the entrance.

Do not remove all the honey

The weight of the brood box is checked after or during honey collection. If broad honey wreaths can be seen under the queen excluder, the food supply would be sufficient for several weeks. Usually the edge combs of the second brood box are also filled with honey. These are usually left to the colony or only taken away if it concerns honeydew honey, flow is no longer expected and the feeding for winter is approaching.

If a honey bee colony is found to be (too) light during the weight check, at least one full honeycomb must remain in the honey box, so that the colony does not have to go hungry in bad weather or missing flow. The colonies usually breed at a high level in the summer, regardless of whether there is flow or not. They consume food and need food for breeding. They should not stop breeding for lack of food. Any forced breeding restriction, even in a low-flow summer, has a negative effect on the health of the colonies. If you save on food, you save wrong!

After the honey harvest, a box as honey box is filled with empty combs and foundations. If further good flow is in prospect, the second honey box with empty combs and foundations can also remain on the colony. It is better to wait until the first honey box has been filled again before setting it up. That happens occasionally. Usually the fourth box is no longer necessary.

Ten full honeycombs contain approximately 25 kg of honey. One can be very satisfied if more than 20 honeycombs can be taken from a colony during a season.

Some things to know about honey

The honey bee owes its name to honey. But not only honey can be harvested from her. She also supplies other products such as pollen grains, wax, propolis, Gelée Royale (royal jelly) and bee venom. It is beyond the scope of this book to discuss all these products in detail. Let's limit to some things about honey.

Honey is produced either from nectar or from honeydew. Instead of "or" the introductory sentence could also contain "and". The bees of a colony often collect nectar and honeydew until the honey harvest is scheduled.

If the nectar share is clearly predominant in the honey, it is referred to as blossom honey. If it clearly contains more honeydew than nectar, honeydew honey will run out of the extracter. If the "clear" is not present in both cases, it is a mixed honey from nectar and honeydew.

Nectar is not exactly nectar and honeydew is not exactly honeydew. It just depends on which plant it comes from (see also the table here after).

Varieties of honey are easier to distinguish than mixed honeys. Varietal and mixed honeys are distinguished by their composition and therefore also in taste. Moreover, consumers are different. Children like sweeter or sweet-tasting blossom honey. Many adults prefer honeydew honeys. the spicy, dark pine honey is particularly popular. But it is not available every year. If you can taste desert flower honey, by the way, you should try it once in your life if you can purchase it. Further it is told that New Zealand honey is medicinal, thanks to all the polyphenols it contains.

Popular with some beekeepers is the placement of hives on the heath. The heather honey is often eagerly bought. However, it is difficult to extract. You then need a kolb device or a special (uncapping) roller to loosen the honey in the comb before extracting. And even then it is not possible to get all the honey out of the combs. A solution is to divide the frames for honey storage by the honey bees with slats and then, when the compartments are filled with heather honey, cut them out.
The heather honey is then sold as comb honey or extracted with a press. Really something for the enthusiast.

The first condition for pine growth is the massive occurrence of the green pine honey aphid, which does not occur every year. If the honeydew producers are absent on the pine trees, there is no pine honey. This also applies to the spruce, which is home to several honeydew producers that are important for honey bee productivity. You can count on a good spruce crop every second or third year. A pine crop occurs even more rarely. And if honey comes from the pine trees, it doesn't happen everywhere. If you want to harvest honeydew honey, it is not enough to place your bee colony somewhere in a forest. One must first ensure that honeydew production will also occur.

The proportions are simpler for the blossom or flower arrangements. Meadows, fruit, rapeseed, chestnut, phacelia, linden and sunflowers bloom every year, although not every year equally well. In addition, nectar secretion is dependent of the weather and the location.

The sunflower behaves in a particularly moody manner. Although this is intensively flown by the bees, the yield for the beekeeper is usually very modest. Once honey bee hives of a beekeepers club in the western part of Europe were placed beside fields of sunflowers. The harvested honey was nice, but most certainly it was no sunflower honey. That it is different elsewhere can be found out in the southern countries if one asks the native beekeepers there about the honey yield from sunflowers.

The ingredients of honey

The main components of honey are sugar and water. The water content of mature honey is between 15 and 18%, which is many times less than in the raw material. About 80% of the honey consists of sugar. Sugar usually refers to cane sugar or beet sugar, which can be bought in the store. The cane sugar also bears the name sucrose. It is a double sugar composed of the simple sugars glucose and fructose. Glucose and fructose are also known as grape sugar or fruit sugar.

<table>
<tr><th colspan="5">Sugar contents</th></tr>
<tr><td>of:</td><td>% sugar</td><td>% saccharose</td><td>% glucose</td><td>% fructose</td></tr>
<tr><td></td><td></td><td></td><td>(grape sugar)</td><td>(fruit sugar)</td></tr>
<tr><td>phloem sap</td><td>5 - 20</td><td rowspan="9" colspan="3">Most of the sugary entities mentioned have an enormous variety with a diversity of sugar values for each variant. The further elaboration of these ranges of contents is generally too extensive for this table. For further interest, reference is made to the flood of publcations on the subject. The aim here is to give an impression of the immense variation in sugar contents of already the most well-known nectar plants or their products.
Nectar contains an amount of water of about 80%. In general, for every 100 grams of nectar, honey bees need to evaporate 75 grams of water. This leaves about 25 grams of honey with less that 5 grams of water. An amazing achievement indeed.</td></tr>
<tr><td>honey dew</td><td>4 - 8</td></tr>
<tr><td>nectar</td><td>3 - 80</td></tr>
<tr><td>honey</td><td>80 - 85</td></tr>
<tr><td>plum</td><td>12.90</td></tr>
<tr><td>apple</td><td>21.30</td></tr>
<tr><td>cherry</td><td>35.10</td></tr>
<tr><td>rapeseed</td><td>45.10</td></tr>
<tr><td>dead-nettle</td><td>42.40</td></tr>
<tr><td>linden</td><td>29.50</td><td>73.70</td><td>13.70</td><td>12.60</td></tr>
<tr><td>black berry</td><td></td><td>28.00</td><td>37.90</td><td>33.90</td></tr>
<tr><td>red clover</td><td></td><td>62.80</td><td>10.10</td><td>27.10</td></tr>
<tr><td>yellow mustard</td><td></td><td>0,00</td><td>53.60</td><td>46.30</td></tr>
</table>

Sucrose can also be found in honey, but it only accounts for a few percent of the total sugars. Glucose and fructose predominate in honey. Nectar and honeydew clearly contain more sucrose than honey. When the collected raw material is thickened, the sucrose is converted into its components by an enzyme from the honey bee saliva. In addition to sucrose, nectar and honeydew also contain glucose and fructose and many other types of sugar, simple, double, triple and multiple sugars.

The best known triple sugar is melecitose. It consists of two molecules of glucose and one molecule of fructose and can only be found in honeydew as it is synthesised by the aphids themselves. The content varies between 10 and 70%. A melecitose-rich honey saccharifies very quickly. If the bees only ingest honeydew with a high melecitose content, it may happen that the melecitose content is not reduced enough during conversion to honey and the honey in the combs becomes saccharified and solid before harvest. This "cement honey" cannot be extracted.

Combs with "cement honey" can serve as food frames for the young colonies in the summer. It is not suitable as a winter supply in the honey bee colony. The bees have to bring in water to dissolve it. In addition, it contains a lot of ballast substances that would unnecessarily burden the excrement bladder after digestion while the honey bees have less opportunity in winter to empty their excrement bladder.

All honey chrystallises, even if it does not contain melecitose. Typically, it only solidifies after being extracted and stored in a bucket or glass jars. Depending on the type of honey, it takes weeks or months for the honey to saccharify. The crystallisation is determined according to the glucose-fructose ratio. The higher the grape sugar content, the faster the honey crystallises.

Rapeseed honey and dandelion honey have a high grape sugar content. These honeys can already become solid in the comb if you wait too long after harvesting before you start extracting. Than you have the same problem as with heather honey.

In honeydew honeys, the fructose predominates. These have the necessary time before they start to crystalise, provided they contain little melecitose. This is usually the case with pine honey. Pure pine honey remains liquid for months after extracting.

When honey chrystalises, it becomes less attractive and sometimes also rock hard. A beekeeper must take this into account when selling. Many honey customers refer to the crystallisation of honey that they have purchased in liquid form as "sugaring out" and suspect that the beekeeper has mixed the honey with sugar. In such cases, information about the correct handling of the honey can prevent misconceptions. It can be put on the jar label that after a while the honey will crystalise being a sign of its quality and in that case one can warm the honey up till 38°C without destroying the quality and all the good ingredients. However, in principle honey should be sold in the condition in which it is consumed.

The great diversity of honey is not or not only determined by the large number of different sugars. Honey also contains minerals, organic acids, including formic acid, lactic acid, oxalic acid and citric acids, amino acids, vitamins, enzymes or ferments and aroma substances.

These and other substances can also occur in the formed ingredients of honey, in pollen grains, mold spores and hyphae. Any summary of the chemical composition of honey would remain incomplete, as not all of it would be complete components of honey are known. This is no wonder given the great diversity of honey.

Honey characteristics

- consistency
- colour
- smell and taste
- physical properties
  electrical conductivity
- chemical composition
  water
  sugar
  nitrogen compounds
  ferments (enzymes)
  minerals
  acids
  aromatic substances
  vitamins
  unknown matter
- formed ingredients
  pollen
  mold spores and threads

The diverse composition of honey justifies its special value. Honey is therefore more than sugar. That's right. Is honey therefore healthy or promotes health? One can argue about that.

Fruit and vegetables contain substantially more vitamins, organic acids and minerals than honey. These and other ingredients only occur in trace amounts in honey. Maybe it is precisely that which is the cause of its health value.

Anyone who substantiates the value of honey by the fact that honey contains a variety of natural substances that are also active when they are only present in traces, cannot dismiss the appearance of foreign substances in honey by saying that they are harmless because they are only present in small quantities occur. Such an argument is contradictory and therefore implausible.

Pure honey from clean combs

Honey has a good name. It is used in folk medicine. Its high prestige is linked to the claim of freedom from foreign substances and environmental poisons. The greatest danger to honey is the treatment of honey bee colonies against diseases, in particular Varroosis and malignant foulbrood. Antibiotics and varroazide have no place in honey. The honey will most certainly be free of these substances if no residue-forming agents are used at all.

In addition, honey is not only at risk from the direct introduction of harmful substances. That would certainly happen if the honey bee colonies were treated untimely, during flow. However, the residues can also take the detour via the wax and end up from the wax into the honey when it is stored in the combs. Even when sieved and skimmed, each honey contains fine wax particles. If these wax particles are contaminated, the honey will contain residues, even if only in small traces.

As far as residues are concerned, honey quality is decisively determined by the purity of the wax. Anyone who wants to keep their honey free of residues should not use products that can enrich the wax to combat varroa and should also refrain from purchasing foreign wax in the form of foundations. It makes no sense to treat the colonies exclusively with organic acids to avoid residues in the wash and then to purchase these residues in the form of contaminated foundations.

Anyone who wants to act consistently should only use proprietary wax. Then one knows exactly what one has. In a well-operated beekeeping operation, sufficient wax is supplied in the form of old combs, drone combs and sealing wax. After melting, this is then used to make center walls and candles.

A water-cooled center wall casting mold is expensive. the purchase only pays off when there are large numbers of colonies. One possibility would be for the beekeeping association to purchase a mold and rent it to members for a small fee. This is practiced in many associations.

Cold preserved

Nectar and honeydew have such a high water content that they would spoil in a short time at the temperatures prevailing in the colony. This is prevented by the bees by rapidly thickening the collected raw materials. The communal achievement necessary for this is one of the many admirable skills of the honey bee colony.

The collected substances are constantly in motion in the honey bee colony. The workers pass their honey bladder contents to the hive bees or store them at the bottom of the brood box in empty cells near the entrance. It does not remain there for long, but is absorbed by the hive bees, transferred to others, stored again in the meantime, absorbed, transferred and thus slowly moved upwards into the honey box. In this feeding chain the collected substance becomes slowly thickened. Water evaporation is achieved by continuous ventilation.

The hive bee spreads the collected substance over and over again on its outstretched tongue. During this surface treatment, the water evaporates, the collected material dries and becomes increasingly liquid. During the thickening process, some honey bee saliva is added to the honey in the making, which contains enzymes that change the sugar spectrum and thereby increase the proportion of simple sugars. This process is of great importance.

An 80% sugar solution only remains liquid if it consists largely of simple sugars. Cane sugar would crystallise before this degree of thickening was reached. The bees have difficulty using the crystallised honey in the winter. On the other hand, reducing the water content to 18% is absolutely necessary to preserve the honey without spoilage.

At the high temperatures that prevail in the floor above the brood nest, fungi would multiply quickly in a water-rich honey and cause it to ferment. Continuous air freshening is also necessary for water evaporation. During the ventilation in which the entire population participates at night, warm hive air is exchanged with cool fresh air.

The honey preparation by the honey bees in brief

- Introducing nectar or honeydew into the hive
- Transfer to the hive bees
  => feeding chain,
  repeated ventilation,
  addition of endogenous ferments (invertase, glucose oxidase)
  => thickening
- Storage of the honey in the comb cells, fill up gradually
- Water evaporation through continuous heating and ventilation (airing)
- Sealing of the filled cells with wax

Simultaneous heating heats the fresh air so that it can absorb more water. This evaporation of water from the honey functions very well. A colony can release more than 1 litre of water in one night in this way if it has previously consumed a correspondingly large amount of thin liquid nectar or honeydew.

When the honey cells are filled with ripe honey, they are sealed. The wax seals prevent the honey from attracting water. Honey is more than happy to do that when it is open to something. Therefore, the honey should be stored tightly closed in a glass jar or bucket.

Harvest and storage of honey

Honey is best preserved in a honey bee colony. As soon as it is removed from there, further processing and storage is at stake. Even with careful handling, the honey does not have an indefinite shelf life. Its shelf life is primarily determined by its water content and storage conditions. The cooler and drier the storage area is, the longer the honey will last. A constant temperature of approximately +10 °C and a relative humidity of less than 50% would be ideal. But not every beekeeper can afford an air-conditioned cold room, most of them only have less comfortable spaces available. Often one has to make do with a "cool" room in the basement. The warmer or more humid the storage area is, the faster the honey loses the quality it had when it was harvested from the honeycombs.

Compared to honeydew honeys, blossom honeys are more sensitive to unfavourable storage situations. That is why it is commonly tried to store flower honey for a year

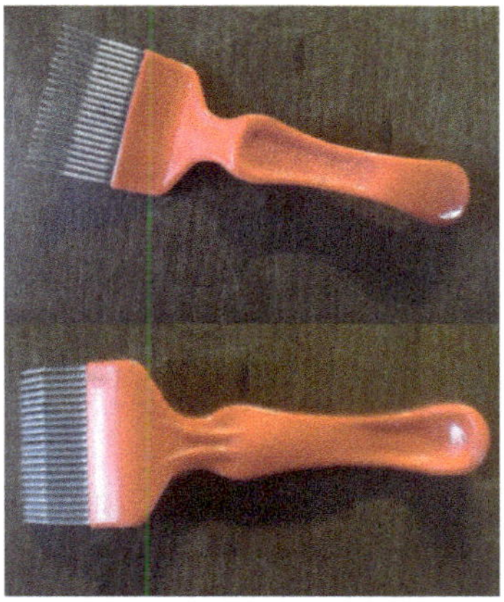

Traditionally, an uncapping fork or -knife is used to remove the sealing wax from the honey-filled comb cells. This causes a terrible mess and requires a lot of cleaning up after extracting the honey. The alternative, which makes unsealing a very easy and fairly clean job, is to use a hair dryer. With the hairdryer, the sealing wax melts to a very small volume that remains on the honeycomb. No seal wax collection tray is needed. After unsealing, the honeycomb can be put directly into the honey extractor. The loss of sealing wax, for example for making propolis ointment, does not outweigh the convenience of using a hairdryer or similar device with a hot air outflow.

at most. As soon as the new harvest has arrived, the rest of the old honey supply is fed to the young colonies. This quantity to store is limited because normally only as much honey is sold as can directly be market in one year. What remains is not even extracted at all, but remains in the combs.

Honeydew honeys last considerably longer without noticeably deteriorating in quality.

However, it is not there every year.

The biggest danger to honey in storage is that it starts to ferment. The drier the honey, the less this danger is. However, it is not enough that the water content at harvest is below 18%. That limit only applies if the honey is in a liquid state.

This does not apply to saccharified honey. Crystallissed honey can also enter fermentation if its water content is below 18%. However, it takes a while before this happens, with honeydew honeys significantly longer than with flower honeys.

Harvesting and processing of honey in brief

- collecting the sealed honeycombs
- unseal
- extracting and possibly sifting
- clarify and skim ---------------------------> * stir
  (if you want creamy honey)
  * fill if the honey is stored
  and crystalised:
- store
- reliquefaction (skimming before and after)
- stirring (if you want to make cream honey)
- filling (always use clean glass jars!)

From uncapping to filling

How do I proceed after collecting the honeycombs?

First the sealed honey is uncapped. The classic device for this is the uncapping fork or uncapping knife. It is indispensable and is also needed when using other devices. A significantly easier method is uncapping with hot air, which can be obtained with a powerful hair dryer. The hot air stream is directed onto the sealed comb surface for a few seconds. Between the cell cap and the honey underneath there is a thin air cushion that expands due to the heat supplied, causing the cap to open or even fall away. The opened cap liquefies in a few seconds and forms a small wax droplet that hardens again when cooling and remains attached to the cell wall. The honey in the uncappled cells remains undamaged. The advantages of uncapping with hot air are:

- Uncapping is faster.
- The cells are not damaged, so that fewer wax particles fall when swinging and the double sieve does not become clogged as quickly as when swinging honeycombs that have been uncapped with an uncapping fork or uncapping knife.
- There is no need to clean up cap wax.

The last argument can also be seen as a disadvantage because there is no uncapping wax for casting foundations. The next disadvantage is that sometimes still the washing caps that fall away contaminate the workplace and its surroundings. That can be overcome by protecting the workplace with suitably cut cardboard or just a strip of cellophane on the work surface (the counter, if working in the kitchen) to collect those few wax caps.

Only clear combs are suitable for uncapping with hot air. On incubated combs, the thin air cushion under the wax cap is missing in the filled and sealed cells. Anyone who uses dark combs in the honey bin should definitely reach for the uncapping fork.

When extracting, the honey flows into a clean bucket. Only containers suitable for honey (for human food consumption) should be used. A double sieve can be placed on the bucket to filter the coarser and catches finer wax particles. But it can also be done without a sieve.

After extracting, the honey is left to stand for at least a day or a few days. During this time the honey becomes clear. The wax particles rise and can be scraped away with an appropriate scraper or spoon after clarification. The skimmed honey serves as food.

After skimming, the honey can take two paths, a shorter and a longer one. The longer route first leads to the storage room, while on the shorter route the honey immediately ends up in the glass honey jar. When filling, a filling jug (or filling bucket) with a cutting tap is recommended. After filling the filling jug (sifting) the honey is allowed again some time to clarify. Only then are the glass jars filled.

It is a possibility also to stir blossom honeys before filling in a glass jar. Stirring controls the solidification of the honey and prevents it from crystallising hard and coarsely. Stirred honey has a finer, firm consistency and is easier to spread on sandwiches or rolls. However, cream honey is not very popular with every consumer.

Stirring can be done in the bucket, but also in the filling jug. One waits until the honey starts to saccharify. The saccharification process starts with the honey becoming cloudy. By stirring (with a suitable stirring device) the sugar crystals that form are broken up and finely distributed. Stirring is repeated several times until the honey has become so stiff that it can just be filled. Anyone who misses this moment and wants to fill it too late must first "thaw" the honey again.

If the honey stored in a bucket is going to be sold in the glass, it usually first has to be liquefied again. Only pure pine honeys are still liquid even after several months of storage. This and the forest honeys do not need to be stirred before filling, because the risk of forest and pine honey solidifying in the glass jar before the customer has consumed it is very small.

This is different with blossom honey. For reliquefaction you need either a warm water bath or a heating cabinet. The honey should not overheat when reheating. The thermostat is not set higher than 40 °C so that heating damage is kept to a minimum. Before or only after liquefaction, the honey is skimmed off, possibly strained into the filling jug when pouring, and then treated as described above. If the honey needs to be stirred, it should not be heated until it is clear. Heating is stopped when the honey is ready to be stirred.

However, this can only be done with honeys that have already been perfectly clarified and skimmed before storage.

Legal obligations must be observed when selling honey. Furthermore, the sale of honey is subject to food legislation, the lot number regulation, and the packaging regulations, the calibration system and hygiene regulations for foodstuffs.

The care of bee colonies in late summer and autumn

A bee colony survives the winter if it is healthy and strong, if possible, is housed on fresh combs and enters the winter with a young queen and has sufficient food supplies. Not much needs to be done to bring the colonies into this state. However, the little that needs to be done is the content of the care in late summer and autumn with which each bee year ends and the next begins.

Young colonies originate from the brood layers that were formed in May/June or from the fertilisation colonies. They are cared for differently than the older colonies that were used with the nectar flow in spring and summer for honey production.

The young colonies

Young colonies are usually overwintered in a box. In their case, the late summer care is limited to the varroa treatment after feeding. A varroa treatment before feeding is not necessary if the brood layers or the fertilisation colonies (they originated during the "population multiplication in four steps with integrated queen breeding") have been treated with oxalic acid in the brood-free state. Only the strongest young colonies are placed more spaciously for the feeding to be carried out in August.

The box for the larger setting contains built-out frames and foundations. In such a box, few brooded empty combs and few brooded "pollen frames" can also be hung, which were previously obtained when setting older colonies smaller. The strength of the young colonies is assessed again in October when they have formed a dense cluster of bees during a cold period or after a chilly night. This cluster of bees must then occupy at least four comb seams. The colonies that are assessed as too weak are united by simply stacking them. The queen is caught beforehand, which can be used to replace the queen in an older colony. This procedure can also be carried out in principle when replacing the queens is carried out in October. Ideally, twice as many young colonies are then formed as there are older colonies. This is easy to achieve with the "four-step colony propagation with integrated queen breeding" as explained earlier already. Then two young colonies sitting in one box are united and one of the two queens is used for the exchange of the queen of an older colony. After the queens have been exchanged in the older colonies, these are put together with many young colonies on two boxes for the wintering. The number of colonies then doubles every year if no colonies fall out or are reorganised.

The perennial colonies

There are two concepts for the care in late summer of perennial colonies. The older one is the "complete formic acid treatment" concept, which involves "late summer care in four steps", namely "reduce - treat - feed - treat". Both treatments before and after feeding are carried out with formic acid.

The more recent concept is called "split and treat" or the "SaT" concept. Only oxalic acid is used for this.

Essential components of both concepts are the five guarantees for successful wintering already mentioned:

- varroa control ("healthy")
- rehabilitation and reunification of weak colonies ("strong")
- comb hygiene ("on combs that have been built out as fresh as possible")
- (winter) feeding ("sufficient food supply") and
- exchanging queens from older colonies whereby the old queens are replaced by a younger one before they become too old.

In both concepts, the guarantees for each concept are implemented specifically. That is why each concept is first presented in principle and then described in detail in its working method.

The "Split and Treat"-"SaT" concept

In the "SaT" concept, a production colony is divided into a split with a queen (queen laying) and a "split with brood and bees" (a brood colony) on Day x and both colonies are subsequently treated with oxalic acid in a brood-free state on Day x+2 and Day x+21 respectively.

These treatments are largely independent of the weather conditions, which makes fixed term planning possible. The "SaT" concept starts immediately after the harvest of the summer honey, usually in July, after the use of forest and pine forage also in

August. The earlier the colonies are split, the more time is available for the colonies to prepare for the winter or the stronger they are during the wintering. The optimal span of time to start is between mid-July and early August.

When splitting the production colony, the split with the queen is formed in the former honey super occupied exclusively with clear combs. For the "brood colony" a second mesh bottom box and a second cover (foil, inner lid and metal roof) are required.

Two days after the split, the split with the queen (its formation presupposes flight weather after the splitting) is treated in a (thus stil) brood-free state with oxalic acid against the varroa mite and than three weeks later the brood-free or still brood-free "brood colony" that is reduced immediately before its treatment against varroa. The lower brood box with the older, dark combs is removed, which achieves comb hygiene.

Usually, the queen of the "brood colony" that emerged via a superceding cell starts laying eggs shortly before or after the reduction. Summer weather is essential for the mating flight. Despite the "drone slaughter" in the summer, there is no shortage of drones until the end of August. If the division of the colony has taken place in July, both colonies usually develop into units capable of overwintering until October, which can then be overwintered in one box.

When setting up and feeding both sub-colonies, it must be considered whether they will be reunited in October. If their reunion is firmly planned, the "breeding colony" is placed on or next to the split with the queen after the division. Both sub-colonies then receive approximately ten litres of syrup or sugar water in one or two feedings after their varroa treatment.

What is a split with a queen and how is a split with a queen formed?

The bees of a colony have flown in at the place where the colony is located. If the colony is moved within the flight radius, this is not registered by the flown in bees. They leave the beehive for the gathering or exploration trip and return to the old place. Because they no longer find their home, they try to fly in at a neighbouring colony before their energy to fly is used up. It also happens that homeless bees form a bee cluster somewhere near the former location.

The moved colony, a few meters is already sufficient, loses a considerable number of flying bees through such an action. This can be about two-thirds of the number of bees in the colony.

When forming a split with a queen, the departing bees are not simply left to their fate. A complete hive is placed at the old location where the colony originally stood: mesh bottom box, brood box, foil, lid and roof. Depending on the time of year and the forage situation, the box is equipped with combs and/or foundations and has at least one feeding frame (as a reserve for bad weather). If the split is formed without a queen, it gets a frame with open youngest brood. A few worker egg cells must contain eggs or just hatched or at most two-day-old larvae. What happens? The returning bees notice that no one is home anymore and above all the queen is missing. The absence of a queen immediately prompts them to make queen cells (superceding cells) with some open brood. When the first queen emerges after twelve days, she tries to kill her younger royal sisters or half-sisters, in which she usually succeeds in doing.

About a week later she is sexually mature and, as soon as the weather permits, she goes on a mating flight. When she returns from there healthy and fertilised, she starts laying eggs. It takes another three weeks until the first workers emerge and the split that has long ceased to be a split as such begins to grow. By then, the number of bees in the developing colony has decreased by more than half.

The collecting diligence of the flying bees has also suffered under the queenless state and the reversal from "outdoor service" to "indoor service". It is rare for a queenless composite split to completely build out the foundations in the hive and fill them with honey, even if there is a good supply of nectar during the queenless and brood-poor period.

The split with a queen acts differently when it is formed with (its) queen. Then its collecting zeal is less paralysed, but still clearly because a large part of the old honey bees have to take care of the growing brood nest. The flying bees do indeed survive longer (they live considerably longer than in an intact colony), but their number nevertheless decreases very quickly, just like the split formed without a queen. A few days after the first brood they cared for hatches, the last old bees with which the split was formed say goodbye to life. Nevertheless, the number of honey bees and also the brood nest size increase again with the outcome of the first generation of bees in the split.

The "SaT" concept in detail

On ***Day x***, the two or more years old colony that has been kept on three boxes (two brood boxes and a honey box) is divided into a split with the queen that is placed in the honey box and into the queenless "brood colony" (a split with brood and bees) that is placed on an additional mesh bottom box and also covered with foil, lid and roof. For this reason, a second mesh bottom box and a second cover (foil, lid and roof) are required for the division of each over-aged colony. The "brood colony" is placed to the side in such a way that the flying bees fly off and return to the split with the queen.
For this purpose, it can be placed away from, to the side of or on top of the split with the queen. If the reunion of both sub-colonies is planned in October, the "brood colony" must be placed on or directly next to the split with the queen. If placed next to the split with the queen, it must be raised by two boxes during the flight phase to imitate the "skyscraper" that the flying bees are used to. The "breeding colony" can also be placed on the split with the queen until reunion and next to it two days after splitting if the varroa treatment of the split with the queen is to be carried out. If not, the "breeding colony" remains on top of the split with the queen until reunion and is taken off each time treatment or feeding of the split with the queen is on the programme and then placed back on top.

On ***Day x+2*** the split with a (the) queen is treated with oxalic acid. Dripping works (most cerainly in summer) significantly worse than spraying*. That is why the spray treatment or to combine both is preferred. In this "sfb-sfa" treatment ("spraying from below - spraying from above"), the honey bee cluster hanging in the mesh bottom box is first sprayed gently from below with oxalic acid solution after tilting the box, which also causes the bees from the cluster to retreat into the comb seams and then sit there more closely together than before. Then the same solution is sprayed into the occupied comb seams from above with a syringe. A pipette end on the syringe ensures a thin jet. A thin jet achieves a better distribution of the solution on the honey bees than by dripping on thick drops. For this treatment of a split with a queen, approximately 80 - 100 ml of oxalic acid solution is used if nine seams are densely occupied.

Dosages of amount of oxalic acid crystals on specific water amounts for spraying

1 colony:	2.4g oxalic acid crystals per 80 ml tap water
5 colonies:	12g oxalic acid crystals per 400 ml tap water
10 colonies:	24g oxalic acid crystals per 800 ml tap water

Pour the solution into a plant sprayer. Make sure you have the wind at your back when spraying. Use adquate protection (gloves, face mask, goggles).

The treatment is carried out early in the morning before the start of flying or in the evening after the end of flying out.

*[In the meantime, the research is going on whether a treatment of the split with the queen really is necessary or can be skipped awaiting the evaporation of oxalic acid in autumn. The evaporation or sublimation of oxalic acid is very effective and is also tolerated very well - and better than spraying - by the bees. The advice for now is still to treat with spraying of oxalic acid. The evaporation of oxalic acid - for example sublimating two grams of oxalic acid crystals with a stove and a tea light - can also be repeated without any problems if necessary. See also "The evaporation of oxalic acid" that will follow later on.]

On ***Day x+21*** the "brood colony" is brood-free if it had no drone eggs at the time of the division 21 days earlier. Otherwise, brood-freedom is not achieved until Day x+24. In the "brood colony" lives a queen from a queen cell that emerged on Day x+12 and that may have already completed her mating flight on Day x+21. It is more likely that her mating flight will not take place until the next period of good weather.

First the "brood colony" is made smaller. The bottom box (with the old combs) is removed. This intervention is best done during air traffic, the subsequent varroa treatment later. In the evening, when there is no more flying or otherwise the next morning before there is air traffic again.

The "brood colony" is then more compact on ten frames than on twenty frames. The treatment is then carried out as ("spraying from below - spraying from above" with oxalic acid) as with the split with the queen three weeks earlier.
If good summer weather prevails on the day of the reduction, it should be borne in mind that the young queen could make her mating flight. She should not be disturbed during this.

The "dividing and treating" also includes the pre- and post-treatment diagnosis. The "brood colony" loses significantly more mites after its treatment than its split with the queen three weeks before. If there are more than 1000 mites, a quick follow-up treatment should be considered, so that not too many "residual mites" multiply in the winter bees of the "brood colony". When planning and carrying out this follow-up treatment, the weather conditions, the colony development (when did the "brood colony" go into brood?), the care for the colony (is the "brood colony" reunited? If so, when? How and when is it fed?) and the development of the infestation (which can be reliably assessed when the first brood starts to hatch) must be taken into consideration. Many variations are possible. The combination of "reunion" in October followed by an "OA block treatment" [repeated Oxalic Acid Vapour treatments] in late autumn/early winter has proven to be effective.

The way to execute the "split and treat"

The division of the population on Day x

1. Remove the (metal) roof and turn it upside down on an empty box that has already been prepared.
2. Remove the honey box (estimate the food supply based on its weight) and put it to one side.
3. Remove the upper brood box (estimate the food supply based on its weight) and place it diagonally on the turned-up roof.
4. Remove the lower brood box from the mesh bottom box and place it on a second mesh bottom box that has already been brought along, the entrance of which has already been reduced with a strip of foam rubber so that the bee cluster hangs as close as possible to the entrance.

5. Place the honey box on the old mesh bottom box, reassess the food supply and, if it is found to be too little, lift an empty edge frame and put it to one side.

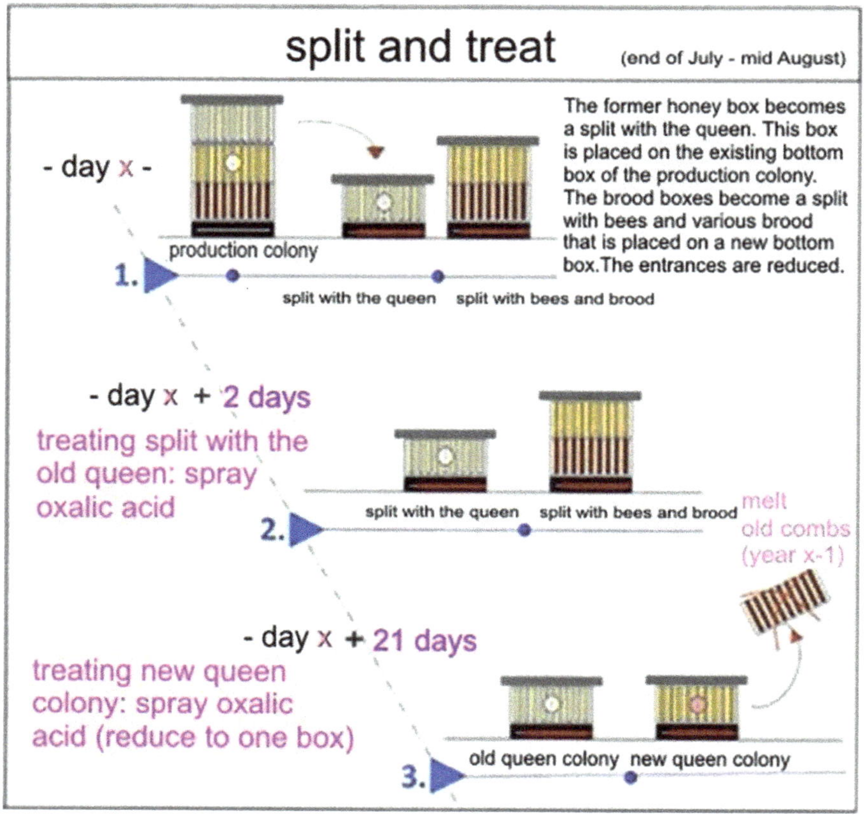

original (German) picture source: Dr. O. Boecking, Celle

6. Search the frames of the upper brood box and then, if necessary, those of the lower brood box for the queen, catch the queen, lock her in a cage closed with a feeding dough together with two or three workers and place this on a comb alley of the honey box; if necessary (see point 5.) exchange a lifted empty frame of the honey box with a rim frame from the upper brood box that is well filled with food.
7. Cover the honey box with foil, inner lid and metal roof.
8. Put the "brood colony" (= split with brood and bees) back together (the upper brood box on the lower brood box), assessing the food supply; cover with foil, inner lid and roof.
9. Strap the "brood colony" with a tension cord and place it on top of the split with a (the) queen and with the entrance in the same direction, loosen the tension cord.

Tips for searching for the queen...

With a probability of about 80%, the queen is in the upper brood chamber. That is why she is searched for there first. If the queen is not discovered immediately at the first look, it is recommended to search all the frames again and hang them in an extra empty box, so that if necessary the inner walls of the brood box can also be carefully checked. During the second search for the queen, one also pays attention to which frame contains only open brood, which is then, if the queen is still not found, hung as an "exile frame" in the split [unfortunately probably not with a queen yet], so that the bees can make queen cells (supersedure cells) on it. Two days later, when one starts treating this split, it should be much less difficult to discover the queen than in the "brood colony". Once discovered and grabbed gently she is then put in her cage and placed in the split, after the "exile frame" previously freed of bees in the split has been exchanged for a brood-free frame.

The treatment of the split with the queen on Day x+2

Research is underway to determine whether treating a split with an old queen (and fly bees) is truly necessary. Further investigation consistently reveals that there are hardly any varroa mites in such a split. Nevertheless, what seems necessary so far follows:

1. Put on acid-resistant gloves and protection goggles, fill the spray bottle with prepared oxalic acid solution, set the atomiser to "spray"; [if necessary] place a pipette tip on the syringe and prepare the syringe.
2. Remove the brood colony from the split with the queen (first secure it with a tension cord) and set it aside when it was put on the split.
3. Slide a sliding inspection board (inspection board) into the mesh bottom tray of the split.
4. Open the split (remove the roof and remove the lid and foil and set it aside).
5. Remove the queen cage.
6. Tilt the brood box of the split with the queen backwards and spray the honey bee cluster hanging in the mesh bottom tray with oxalic acid solution in a few targeted bursts.
7. Wait until a large part of the bee cluster has retreated into the seams, then repeat the spraying with a few targeted bursts.
8. Place the box back on the mesh bottom container, unscrew the spray bottle head and suck up approximately 50 ml of oxalic acid solution from the bottle with the syringe and [if necessary] put the pipette end back on.

9. Spray the comb seams occupied by the bees from above with a thin jet; the more densely occupied comb seams (under which the bee cluster hung) are "run down" twice.
10. Determine the total consumption of the oxalic acid solution based on the number of spray bursts and the amount injected. This consumption should be between 80 and 100 ml.
11. Cover the split with foil, lid and roof.
12. Place the brood colony back on top of the split with the queen, loosen the tension cord.
13. Check the inspection board (inspection board) of the split with the queen whether and where the administration of the oxalic acid solution has ended up on the inspection board.
14. If this has not been done on day x, (also) slide a inspection board (inspection board) into the mesh bottom tray of the "brood colony" and observe the natural mite fall after four-seven days. In the treated split the mite fall is determined after the treatment.

The downsizing and treatment of the "brood colony" on Day x+21

1. Remove the metal roof of the "brood colony" and place it upside down on one of the two empty boxes you brought along.
2. Remove the top box from the "brood colony" and put it aside.
3. Remove the bottom box from the "brood colony" and place it on the upside down roof.
4. Place the top brood box on the mesh bottom box, remove the inner lid and foil and place an empty box on top.
5. Lift all the frames of the bottom brood box, shake the bees sitting on them into the set up empty brood box and hang them bee-free in another [the second] empty box.
6. Remove the empty box again, cover the downsized "brood colony" with foil, lid and roof.
7. In the evening or the next morning, treat the "brood colony" with oxalic acid solution in the same way as the split with the queen on day x+2.

Feeding and reunion

The split with the queen and the "brood colony" must always be provided with sufficient food. The feeding is easiest to ensure if the former colony had sufficient supplies before the division in both the honey super and the upper brood box, which is guaranteed if each sub-colony has two well-filled edge frames.

The split with the queen can be fed immediately after its varroa treatment. It is most convenient to do this feeding (and the next) together with the varroa fall diagnosis. Since the mite fall caused by the oxalic acid treatment continues for days and the maximum only occurs on the second or third day, one must wait at least three days before feeding [evaporation of oxalic acid can be carried out...].

For [being able to] feed, an empty box is placed on the split with the queen in which a feeding tray is placed or two empty frames are lifted and then a plastic feeding frame box (which is usually two frames wide) is hung on the edge. At the same time, the entrance is reduced with a strip of foam rubber in such a way that it is open two fingers wide on the other side [from where the feeding tray hangs].

With the "breeding colony", feeding is only worthwhile when the queen from the brood cell is laying eggs. This is usually only the case after the reduction or after the varroa treatment. The check to be carried out then or before feeding whether the colony is good for the queen can be linked to a crash diagnosis.

If reunification is firmly planned, each sub-colony receives approximately 10 litres of syrup [or sugar water] in two or three portions after its varroa treatment.

Reunification in October takes place by simply placing them on top of each other. Usually the queen placed on top, i.e. the younger queen of the "brood colony" that broods more intensively and longer in the autumn, survives if it is placed on top of the split with the queen. However, in the spring at the beginning of the swarming period a change of boxes must take place so that the younger combs of the split with the queen come to the top.

Whoever wants to be sure that the queen of choice, the younger of the "brood colony" or the older of the split with the queen, stays alive, catches the other one, or better still catches both queens and places the queen of choice, after she has [possibly] been marked and imprisoned in a cage closed with food dough, on a comb alley of the united colony.

If one takes the trouble, this intervention can also ensure that the clear frames of the split with the queen are already coming up. In addition, the food supply is estimated.

If necessary, a feeding frame box is hung in the upper box of the reunited colony.

What to look for in the "SaT" concept

- The old colony must be in the same place for at least seven days before the division. Only then will it provide a strong split with the queen.
- The honey super must not be set up (again) just before the division, and especially not if it contains freshly spun combs. That invites robbery.
- The "brood colony" usually loses the door guard with the flying bees; therefore the entrance must be made smaller unconditionally.
- The "brood colony" is put back together as it was after the search for the queen. Anyone who already swaps the brood chambers during the division (the lower one to the top, the upper one to the bottom), in order to facilitate the later reduction, ensures that the remaining food is carried upwards and stored in the older combs.

The "Complete Formic Acid Treatment" Procedure

In this procedure, the older colonies are treated twice with formic acid against the varroa mite, the first time before and the second time after feeding. Beforehand, the colonies that have been kept on three boxes (two brood boxes and a honey box above a queen excluder) until mid-August after the harvest of the summer honey are placed smaller on two boxes, whereby the lower box with the older combs, which are usually (only) brood-free from mid-August, is removed.

Colonies that are found to be weak or too weak when they are reduced in size are further cared for as colonies-on-one-box and are cleaned up after the first formic acid treatment or are fed first and united with another colony in the autumn.

The major disadvantage of the formic acid treatment is its dependence on the weather. It must be warm enough during the treatment. The treatment is most successful when it is clearly warmer than 20°C during the day. In shady and damp places it must be even warmer. This dependence on the weather requires the beekeeper to have a flexible time schedule, especially if it is always cold or always autumnally cold for a long time in August and September.

Production colonies are kept in three or even four boxes in the summer. The lower brood space consisting of two boxes and the honey box above it are separated by a queen excluder so that the colony can place its brood exclusively on the combs of the brood space from March to August. Only clear combs are used in the honey box (the super).

From the end of July, the brood nest of the older colonies shrinks very quickly. Because the colony always builds its brood nest directly under the honey wreath, the ever-shrinking brood nest moves upwards during the late summer with its poor nectar flow. The lower brood box usually becomes brood-free from August onwards and can then be removed completely with its older or darker combs. The upper brood box moves downwards, the honey chamber with the virgin combs now becomes the upper brood box. If this reduction in size is practiced every year, the combs in the brood box will not be older than two years. Anyone who practices comb hygiene prevents diseases. Prevention is better than cure, also in beekeeping!

The way to reduce the size

For reducing the size, the colony that has been on three boxes since the summer honey harvest is separated. When separating, the strength of the colony is assessed. This is possible without lifting frames. When removing each box, the number of seams occupied by honey bees is taken into account and before removing the bottom box, a quick check is made to see if and where a honey bee cluster is hanging in the mesh bottom box. Above the honey bee cluster, the comb seams are particularly densely occupied with bees.

The combs in the centre of the honey bee box would also be brooded towards the bottom. As a check, a small frame is lifted from the bottom brood box and set aside. Then, the frames hanging next to it are moved en bloc [as a whole] with the hive tool to check the suspected comb seams for the presence of sealed brood. The result determines the further progress.

The procedure in detail:

1. Remove the metal roof and place it upside down on an empty box placed behind the colony.
2. Put the inner lid to one side.
3. Remove the honey box and place it to one side with the front on the ground.
4. Remove the queen excluder [and place it against the lid, for example].
5. Remove the upper brood box and place it to one side with the front on the ground; before removing it, look at the size of the brood nest from below in the

comb seams of the tilted box.

6. Remove the lower brood box and place it on the upside down metal roof. Before and during this action, assess the bee occupancy of the lower brood box. All comb seams should be occupied with honey bees when viewed from above. A cluster of bees should hang in the mesh bottom box. If this cluster of honey bees is completely absent, the colony is placed on a smaller box! (The further procedure in such cases is described further on.)
7. Place the upper brood box on the mesh bottom box of the colony.
8. If one wants to make sure that there is no more brood in the lower brood box, then all honey bees in the relevant comb seams are driven down with some smoke. If sealed brood can then be seen, a frame on the edge of the lower box is lifted and put aside. The neighbouring frames are then moved en bloc [as a whole] so that the relevant frames can be checked for the presence of brood. The number of brood cells is determined.
9. Lift as many brood-free (edge) frames in the upper brood box placed on the mesh bottom box as the lower brood box has brood frames and put these aside.
10. Lift the brood frames from the (former) lower brood box and hang them on the edge of the (former) upper brood box.
11. Place the honey super on (former) upper brood box, possibly swap edge frames with food from (former) upper brood super with brood frames with empty honey super frames; collect the clear honey super frames that are left over bee-free in a box.
12. Place empty box and shake bees from the remaining frames of the (former) lower brood super into the colony; collect the remaining dark brood frames in an extra box.
13. Shake the bees that have fallen on the roof into the colony, place [if possible, foil and] the lid on the empty box that is placed on the now smaller colony, fold the foil [if it could not also be placed on the empty box] and deposit it on the lid under the roof.
14. If necessary, reduce the entrance.
15. To diagnose the number of mites that have fallen, slide the inspection board into the mesh bottom box.

What to do with weak, over-year-old colonies?

When reducing in August, each colony is already assessed when it is taken apart to determine whether it is suitable for overwintering or not.

When removing the honey box, it is examined from above and below and the bee population is assessed, also in comparison with the first look at the combs of the upper brood box when removing the queen excluder and with the second look in the bottom box when tilting or removing the lower brood box. If only a small cluster of bees hangs in the bottom box or is even completely absent, then usually the comb seams on the edges in the bottom brood box and the honey box are also occupied by only a few bees. This leads to the decision to reduce the colony in question to one box.

Immediately after the colony has been separated (steps 1. - 6. when reducing the size) the following is done:

1. the upper brood box is placed on the mesh bottom box and then
2. an empty box is placed on top of it,
3. all frames of the lower brood box are lifted and the bees sitting on them are shaken off into the empty box that has been placed on top of the former upper brood box, the frames are collected bee-free in a box,
4. all frames of the former honey box are lifted and the bees sitting on them are shaken off in front of the entrance, the frames are then collected in an extra box,
5. the colony is covered with [foil,] lid and roof, [if necessary] the foil is folded up and placed on the lid and stored under the roof.

When reducing, a sufficient food supply must be ensured. This is already the case if there is a narrow honeycomb above the brood nest and both edge frames are halfway filled with honey.

After a diagnosis of the fall, a treatment with formic acid (or oxalic acid) against varroa mites follows. The colony is then fed and the varroa treatment is repeated if necessary. In October, it is united with a suitable young colony. Before the uniting (by placing the young colony on top), by preference the queen of the older colony is removed.

The next step: the first treatment with formic acid

The use of chemicals in beekeeping seems officially not yet allowed in the UK. Nevertheless it is of value (in this age of internet as well) to publish the successful ways varroa is fought on the European continent using chemicals that are also familiar to bees and can be find in honey too. There are no other varroa control agents with so many applications with formic acid. These are discussed in the chapter "Varroosis" later on. Here it is limited for now to describe the form of application that in experiences from numerous tests over many years has proven to be the most suitable for treatment in late summer: treatment from above with the medicinal dropper bottle as in the "Liebig dispenser". This two-time formic acid treatment (once before and once after feeding) also includes monitoring the development of the infestation by means of the precipitation diagnosis before and after the treatments.

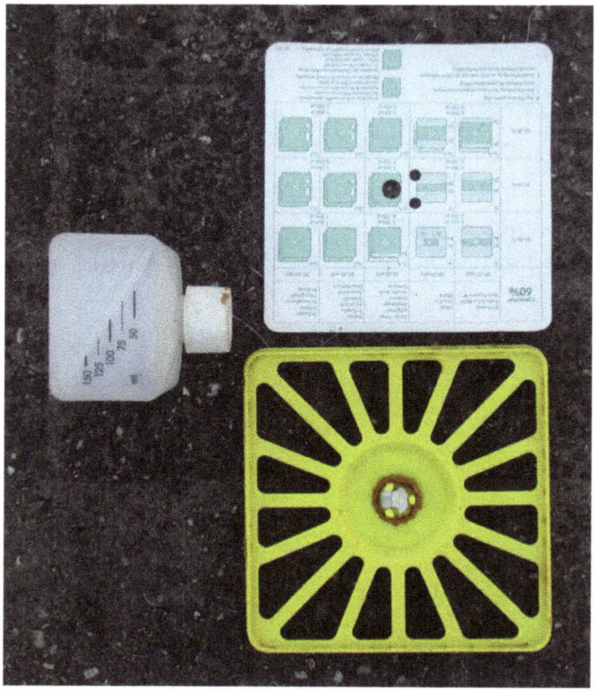

The Liebig dispenser as currently performed by the respective Swiss manufacture. Dispenser bottle, dropper cap, evaporation paper with holder.

Method of treatment with the Liebig dispenser

The square plastic holder has a central saucer, the outer part is slightly raised and broken on the side of the saucer so that the formic acid can evaporate from the evaporation paper upwards and downwards. In the middle of the saucer there are three pointed projections on which the bottle is pressed. A thin and perforated blotting paper serves as evaporation paper. For the treatment of a colony on two boxes at temperatures up to 25 degrees, the paper is used in its entire size of 15 by 15 cm. At higher temperatures, for colonies on one box, for treatments of rear-opening hives or trough hives, for hives with warm construction, the evaporation paper can be reduced as required.

Photo by Dr. G. Liebig

The entrance remains as it is

1. Slide in the inspection board
2. Remove the lid and foil, place an empty box.
3. Place the dispenser with the inserted evaporation paper on top of the colony.
4. Unscrew the dispenser bottle (filled with 100 ml formic acid), making sure that the dropper cap remains on the bottle and does not stick to the screw cap, press the bottle onto the evaporation paper holder, place the screwcap on the bottom of the bottle or on top of the frame carrier of the box.
   Close the empty box with lid and roof. The foil is stored folded on the lid.

After three - four days the dispenser bottle is empty and the evaporation paper is dry or almost dry. Then these aids are removed (wear also gloves when using damp evaporation paper!), the inspection board is removed and the mite fall caused by the treatment is observed and assessed.

It should be remembered that the formic acid is also active in the sealed brood. The mites that are killed there only fall when the treated brood emerges. After the bottle has emptied or after the evaporation paper has dried, it takes twelve days before natural mite fall can be seen again.

The success of the treatment can be assessed by comparing the natural mite fall after the treatment with the natural mite fall before the treatment. In addition, the number of mites that have been controlled can also be counted. For this purpose, the inspection board must be slid in for about two weeks; in total four weeks:

- one week before the treatment,
- two weeks during and after the treatment and
- a fourth week after the weight loss due to the action in the sealed brood.

The checks of the inspection boards can be done at the same time as the feeding.

The following step: feeding

When feeding (with syrup or sugar water), it is taken into account whether it concerns wintering with a box with 10 frames or with two boxes with 20 frames.

Due to the limited frame surface, colonies on one box cannot store more than about 15 kg of stock. This corresponds to 15 litres of syrup or 20 litres of 3:2 sugar water (3 parts sugar to 2 parts water). This quantity cannot be given all at once, but must be divided into two or (better) three feedings in such a way that the brood activity in the autumn is not restricted.

The first feeding is given around mid-August, the second and third are only given to the colonies before or after mid-September when the brood nest is shrinking, so that the food can be stored centrally and the shrinking brood nest is pushed forward or downwards. There is then no need to worry that the colony or its bee cluster will lose contact with the food in the winter.

With colonies on two boxes, feeding is not a problem at all. They obtain at least 20 kg of sugar (as syrup or as sugar water) in one, two, three or four feedings. With them, feeding can be done as early as the end of August if only one feeding is given.

The number of feedings is also determined by the size of the feeding box used.

The second treatment with formic acid

A fall diagnosis is made during feeding, with which the varroa fall is assessed. After the feeding has ended, the old colonies are treated a second time with formic acid during warm weather. An alternative treatment would be with oxalic acid, which is particularly suitable when it is too cold for a treatment with formic acid. Than the colony should be "brood free".

The concepts of late summer and autumn care in brief

What do I do?	***What do I not do?***
Production colonies according to the "SaT concept":	do not feed
mite fall diagnosis from mid-July	do not treat with agents that can dissolve in fat
divide colonies in a split with the queen in the honey box and a split with brood and bees, the "brood colony"	do not feed with feed dough
treat the split with th queen with oxalic acid 2 days later	
treat the brood colony with oxalic acid 21 days later	
feed each split with 10 litres of syrup after treatment	
reunion in October. If both splits are than colonies with over 5000 bees it can be decided to maintain the two colonies as such	
block treatment with oxalic acid evaporated from above	
Production colonies according to the formic acid treatment:	
natural mite fall diagnosis from mid-August	
reduce, removing the bottom brood box	
treat w th formic acid for a short time before feeding (preferably in a medicinal dropper bottle on blotting paper)	
give 1 x 20 litres of syrup [or sugar water]	
treat w th formic acid after feeding [with syrup or sugar water] (preferably in a medicinal dropper bottle on blotting paper)	
recommended: change the old queen in October for a young queen	
block treatment with oxalic acid evaporate from above	

Young colonies:	
natural mite fall diagnosis from mid-August	do not increase young colonies
give 2 x 7.5 litres syrup [or sugar water]	do not place 'splits' on older colonies
after feeding [with syrup or sugar water] a treatment with formic acid (preferably in a medicinal dropper bottle on blotting paper)	
unite young colonies [that have not developed sufficiently], use young queen for exchange with an old queen from an over-aged colony	
block treatment with oxalic acid evaporate from above	

Caring for bee colonies in autumn and winter

Checking the food supply

To check the food supply, it is sufficient to determine the "relative" weight by lifting the hive that most of the time now exists out of one box on a mesh bottom box. The "felt" weight is calibrated by estimating the food supply of the lightest colony in flying weather using the "eighths method" [= Lieberfelder Schätzungsmethode]. For this purpose, frames must be lifted. On this occasion, the population size can also be assessed relatively accurately and the result of this "population estimate" can be compared with the number of occupied comb seams after a cool night.

The colonies can also be weighed with a digital suitcase scale. For this purpose, a hammer handle can be inserted from behind into the opening for the inspection board and/or from the front into the entrance. The hook of the suitcase scale is hung on this or else just at the entrance and/or back of the mesh bottom box. The two "half weights" obtained in this way are added together. It is also sufficient to weigh only from behind (or from the front, if you would like to do so) and to double the indicated value. The colonies must then be arranged in such a way that they are "weighable".

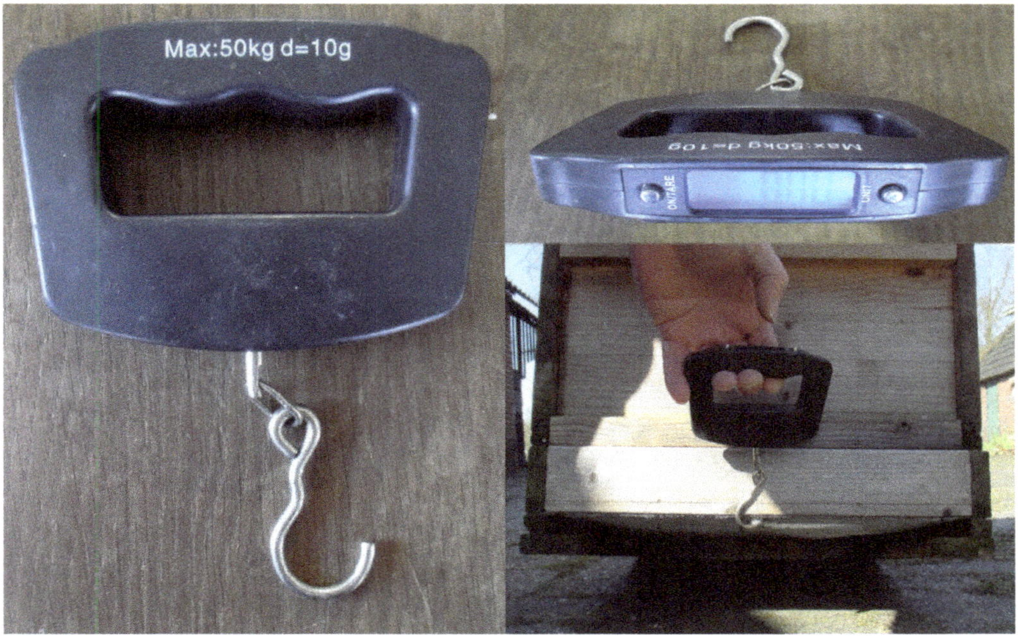

A digital suitcase scale (again). This device that can be obtained for little money enables to determine the easy way the weight of a colony/a colony's food supplies. It is operated "from the back". For knowing the total weight the indicated number of kg has to be multiplied with two.

With the inner lid and the metal roof, the colonies on one box should weigh around 28 kg (respectively 14 kg from the back) at the beginning to mid-October, colonies on two boxes should weigh around 42 kg (respectively 21 kg from the back).

If more feeding is necessary, feeding is either done "from above" (the feeding box is placed in an empty box) or "from the side". For this purpose, a feeding frame is hung next to the colony on the side of the box. Depending on the width of the feeding frame, one or two frames must be removed and hung in an empty box placed above for the time being. This empty box is also used in late autumn or early winter for a possible OAV treatment (Oxalic Acid Vapour Treatment).

If the bees do not eat the food, the food provided must be given closer to the bees. It can also be given "from below" if the entrance is kept small. For this purpose a flat dish (larger ones can hold three litres) is placed at the back of the mesh bottom box (as far away from the entrance as possible) and filled with syrup or sugar water. Nothing may spill or run over the edge, even if the absolutely necessary swimming aids are then placed in it (corks, broken twigs, dried leaves, grass, etc.).

The final inventory in October

Here we anticipate an event in the care of the colonies in the spring and summer that can lead to a "problem" during the year.

In the spring, with the prevention of swarming, the colony increase also takes place. Splits are made from which young colonies emerge. The number of colonies increases, from year to year; especially when no or only a few colonies are lost in the winter.

A good beekeeper can be recognised by the fact that his/her colonies survive the winter, regardless of how the winter is. Because only colonies that fulfil the guarantees for a successful wintering are wintered.

In October, the strength of the colonies is checked. If necessary, targeted reunification is carried out and in this way "two (or three or four) birds are killed with one stone":

- the number of colonies is reduced
- old queens are replaced by young ones
- weak colonies are "dissolved" or strengthened by association with the association, an adequate feed supply is also ensured.

The weather determines on which day in October you do the final inventory of the apiary. It may be cold, but not rain.

The final inventory is prepared on the basis of the evaluation of the protocols (hive cards) that were drawn up from spring to autumn. The older colonies are selected that have to be provided with a young queen or that have to be dissolved. Which colonies in one box should be united with which other colony? By simply placing them on top of each other or by hanging frames?

The planning should also include where the colonies are located on the apiary. Of the paired colonies, one must remain standing, so that the association does not cause too much trouble.

At the apiary, the colonies are brought up to standard in the row order as they are used to. Their strength is estimated by the number of occupied comb seams, the food supply is assessed by a weight check. This is done either by the "felt weight" that is noticed by simply lifting the hive from behind or that with the "half weight" is determined with a suitcase scale whose hook is attached to the opening for the inspection board of the mesh bottom tray or to a hammer handle that is inserted into the inspection board opening. The result is arrested in the protocol.

Replacing queens and uniting colonies

For the replacement of the queens in the selected colonies, their queens are placed in a cage. To do this, they must first be searched for and found.

Also in the autumn the queens usually stay on the brood combs. The shrunken brood nest is usually in the bottom box and the queen walks around on a brood frame in the centre.

Method:

- The metal roof is removed and turned over on the neighbouring colony or placed on a box that is behind the colony.
- The top box of the colony is removed and placed perpendicular to the metal roof.
- In the top box a frame is lifted on the side and after checking the bees that are on it, placed in front of the hive.
- Then frame by frame is lifted, searched for the queen and hung on the side.

The easiest way to change the queens is to simply do a queen exchange in the fall: catch the "old" ones, place the "new" ones in a cage sealed with sugar dough on a comb seam (or hang them on a stick between comb seams, for example).
Exchanging them earlier (in the months of April/May to August/ September) requires two steps: nine days after removing the queen, the queen cells are broken and the "new" ones are placed in a cage sealed with sugar dough on a comb sseam. The colony is covered with foil and the turned-over inner lid (with the hollow side facing down).
In the photo, the cage sealed with feed dough is still completely closed. For free feeding, the feed chamber at the front of the cage can be opened by sliding the slide on the underside slightly downwards, as can be seen in the empty cage that has been placed on the foil in the photo for demonstration purposes.

Photo Dr. G. Liebig

- If the queen is discovered, the frame in question is placed on the box, the queen is picked up by the wings or the thorax, the queen is let walk into a cage and the cage is closed.
- The frames of the colony are hung back in place, the upper box is placed on top of the lower one, the cage with the queen is placed on top of the bee seat and the colony is closed again.

Then the queen of a young colony is sought, placed in a cage sealed with sugar dough and then the caged "young" is exchanged with the caged "old".

The queenless young colony is united by placing it on another colony with a queen.

Protection against mice and woodpeckers

The entrance is provided with protection against mice, either by attaching (mouse) wire mesh with a mesh width of 7 mm or by a mouse strip that leaves an equal size of space. Mouse wire mesh is preferred because if - for whatever reason - a lot of dead bee droppings accumulate on the mesh bottom during the winter, this can easily block the entrance that has been reduced and narrowed by a mouse strip.

Mouse wire before the entrance in late autumn (before the frost at night!) and winter. You can attach that wire just by a few thumbtacks. Mice do love to nest in the colder times of the year in the hives. If this happens in a hive you can lose your colony.

Mice are everywhere. And woodpeckers are also in many places, but woodpeckers rarely look for food on the apiary. If they do, they cause damage by cutting holes in the beehives. If they are successful, they will come more often in the winter, usually only when there is snow. The woodpecker clings to make holes in the front or (more often) at the back of the box(es) and makes holes from which they use their long beaks to bite for bees in the comb passages that are thus exposed.

A bird net can keep the woodpecker away. It must be hung over the boxes in such a way that the woodpecker cannot reach the box walls and cannot slip past the net. Attaching flapper strips and setting up plastic crows to deter woodpeckers do not work.

top left: a woodpecker has come to visit.

left and above: this is what happens when a mouse has entered the hive and made a nest...

below: mice like to live in a hive as well. Here one has set up a guest room under the metal roof.Also lizards like to house there...

The "remaining mite destruction" with oxalic acid

After the final inventory in October, a mite fall check is carried out. Each colony is given a inspection board and after 1-2 weeks the natural mite fall is determined. Because the colonies have no or only little brood and are also not very active, the inspection boards [inspection boards] remain relatively clean. Even with an incubation period of several weeks, there is little waste, the fallen mites are easy to recognize.

If the natural mite fall is higher than 1 mite per day, a varroa control treatment is absolutely necessary. This is done with oxalic acid. Oxalic acid can be used by spraying the frames occupied by bees, by dripping the comb seams occupied by bees or by evaporating oxalic acid.

Each method has its advantages and disadvantages.

Spraying the frames occupied by bees is very laborious. If the colonies are in two boxes, the upper box must first be removed and put aside. Then a frame is lifted from the lower box and also put aside, and then each frame occupied by bees is lifted and the bees on either side are sprayed with the oxalic acid solution.

After treating the bees in the lower box, an empty box is placed there, in which the frames from the upper box are hung after spraying. The only positive aspect of this method is that you know what you have afterwards, because you can use the procedure to make a population estimate according to the Liebefelder estimation method, in which the brood size can be assessed in addition to the number of bees. The protocol can be comfortly carried out using a dictation machine.

Dripping with oxalic acid is less complicated. It must be freezing cold during the treatment. Opening the colonies will not harm them even if it is colder than -10°C. You can leave out a smoker if the colonies are in a winter cluster when it is cold. The colonies are still under a wide feeding wreath in November/December.

With colonies in one box, it is not difficult to recognise the location of the winter cluster, so that the solution can be dripped directly onto the bees in all occupied comb seams. With colonies in two boxes, it is sometimes a bit more difficult, especially if the winter cluster is in the bottom box. Either you use a flashlight or (better) you tilt the top box. Then the solution can be dripped into and not onto the winter cluster.

You must take your time when dripping. It is better to treat the most densely occupied comb seams twice than once. The more bees are dripped, the better the colony will tolerate the treatment. The oxalic acid is then more easily distributed in the colony.

Depending on the strength, a colony receives 30-50 ml of solution; rather 30 ml if the densely packed winter cluster (at temperatures below or around freezing point) occupies 4 to 5 comb seams; rather 50 ml if it concerns 6 or 7 comb seams.

The effectiveness of the one-off treatment is on average 90% for brood-free colonies, but is clearly lower for brooding colonies.

The colonies should only be dripped once with oxalic acid. Repeated dripping is bad for the colonies! Many bees then become over-acidified and fly away prematurely without being found as corpses on the ground. No dripping of oxalic acid any more after New Year!

The evaporation of oxalic acid is better tolerated by the bees at a distance. This treatment can also be repeated several times in late autumn or early winter without causing increased bee mortality. Just like with spraying or dripping, evaporation also kills only the mites that are on the bees and only completely when the weather is cold or cold-rainy (= no flying weather during the day) and the colony to be treated has little or no brood. Then, with two or three treatments carried out at a distance of 1 - 3 weeks, an effectiveness of more than 95% can be achieved, even if the colonies are still caring for some brood. The risk to the practitioner is considerably higher when evaporating oxalic acid than when spraying or dripping. But here too, it is the method that counts.

Worldwide widespread and also permitted in many countries is the evaporation of oxalic acid on the bottom of the hive. The oxalic acid - usually 2 g of oxalic acid as a "powder" or as a tablet - is placed on a small pan that is pushed through the flight slot on the hive bottom and then heated so that the oxalic acid evaporates in a few minutes. The practitioner protects himself against the vapour that also comes out through a closed flight slot with protective clothing from head to toe. When using an "evaporator" it takes about an hour to treat the ten colonies of an apiary.

Evaporation from above with a tea light and a small stove in an empty container placed on top takes considerably less time and is also less dangerous for the practitioner. Evaporation only starts after the hive has been closed. The practitioner does not stay near the hive or on the stand during the treatment. The protection of the practitioner is guaranteed if acid-resistant gloves are worn when initiating the treatment and when clearing away and cleaning the utensils and the empty container with water (after the last treatment).

Evaporation of oxalic acid in an easy way

By simply tapping a spoon in a vice with a hammer twice, a simple but very effective evaporating stove for evaporating oxalic acid is created.

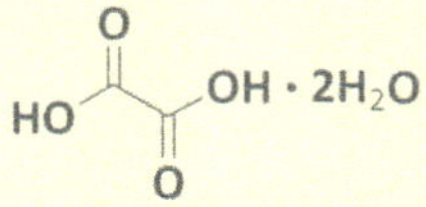

First, after removing the roof and the lid, the foil is removed from the (upper) brood box. Than a piece of structural mesh (mesh for plasterwork) is placed on the (upper) brood box. This prevents the honey bees to walk into the flame of the tea light.

Then the tile (an old bathroom tile, for example) is placed on the mesh. Subsequently the little spoon stove is added. The tile holds the spoon in place and forms the platform for the tea light.

‹ Placing the tea light.

The evaporation device is ready for action ›

*Then about two grams of oxalic acid is deposited on the spoon. Indeed, don't forget to **wear gloves** when doing this.*

Put an empty box on the brood box.

Light the tea light.

Cover immediately the box with the lid. The evaporation will start shortly after. You do not stay near the hive for at least half an hour.

It is even preferred to check the next day whether the oxalic acid crystals have evaporated. If necessary, start the treatment again. Evaporation of oxalic acid is well tolerated by honey bees.

So, in this process you do not need any expensive device and moreover all the sublimation takes place in your absence !!

Check after about five weeks what the number is of the natural mite fall on the inspection board. If necessary so repeat the treatment.

The mite fall caused by the treatment(s) must be observed without further ado and compared with the result of the fall diagnosis carried out previously. When planning the term, it must be noted that the fall of dead mites is increased over a period of 4 to 5 weeks, even if most (approximately 80%) poisoned mites fall in the first week.

The measures that the beekeeper takes in autumn and early winter in brief

What do I do?	What do I not do?
▶ replace queens in old colonies in October and unite young colonies	▶ colonies are not packed. ▶ Honey bees don't warm the hive, they warm the interior of the cluster.
▶ place the mouse lath or (better) the mouse wire in October at the latest (but absolutely not later than the first night with frost)	▶ no treatment with drip agents containing fat-soluble substances
▶ monitor/check the varroa infestation by diagnosing the mite fall in October and November.	▶ no storing of combs that have been brooded (because then you will get problems with wax moths)
▶ treat the colonies with oxalic acid in November/December by dripping (only once, if the colonies are brood-free) or evaporating (repeated	▶ no treatment with dripping oxalic acid after New Year's Eve
▶ treatment is possible and also of still brooding colonies)	
▶ after the oxalic acid treatment is completed, overwinter the colonies above an open mesh bottom box	

When are the colonies brood-free?

Many old colonies leave the brood in October, simply. This is rarely the case with young colonies (and also with the "brood colonies" that have arisen due to the "SaT" concept -split and treat-). This may be because the growth phase of the young colonies only ends in late summer, the queens from the queen cells of the "brood colonies" only develop into brood at a late stage, while the "normally" cared for old colonies have already passed the peak of their development in July.

The brooding process of all colonies in the autumn is influenced by the location and even more by the weather.

The first night frosts in the autumn cause the queens to stop laying eggs. Three weeks later the colonies are brood-free. From then on, the drip method can be used. But this should be done at -5°C rather than at +5°C. At freezing temperatures the colonies are closer together. The closer they are, the better the oxalic acid works when dripping. Opening them during the treatment (removing the lid, possibly tilting the upper tray) does not harm the colonies, even in frost.

When evaporating oxalic acid, you do not first have to pay attention to whether the colonies are free of brood. Nor do you have to pay attention to the weather conditions in the autumn.

If, when checking the mites that have fallen due to the treatment with oxalic acid, clear mite stages are still discovered in the fall, infected brood has emerged after the treatment. In the event of a large mite fall, a repeat treatment is then appropriate.

Winter rest

A brood-free winter rest that lasts as long as possible is the best way to prevent varimorphia (Nosemosis). The cause of this disease lives in the midgut of adult bees and then multiplies when protein is converted in the intestine, which is the case with brood-caring bees. The long-lived winter bees are particularly threatened if they are still active as nurse bees in the autumn or already in the winter.

As long as they do not (have to) care for brood, they limit themselves to an extremely healthy "sugar diet" for them. This is also why the beekeeper must care for his colonies in such a way that the care for brood ends in the autumn and is started again as late as possible. The open mesh bottom box helps with this and also ensures that the unoccupied (edge) combs do not mold during the winter.

In the winter, the highest (and only) commandment applies to the bee population: "Thou shalt not disturb!". Whoever leaves his colony/colonies alone does the least wrong.

The honey bees that are in the winter cluster usually start the new breeding season in February, sometimes even earlier, in January, sometimes later, not until March.

When honey bee colonies start breeding and how intensively depends mainly on the weather, in some cases on the strength of the colony, its pollen supply and its state of health. Breeding colonies consume significantly more food and also lose more bees. In warmer weather they fly away, in colder weather the dead bees accumulate on the ground. The natural bee mortality can amount to an average of around 30 bees per day in the winter. That is around 3000 bees per colony during the winter. That is normal, but it can also be more. Strongly wintered colonies tolerate this loss of bees more easily than weakly wintered colonies.

Colonies that brood in the winter can compensate for bee mortality with the increase in bees that emerge from the brood. This can also lead to colonies growing in the winter. However, this is an exception and only occurs when the winter is very mild. The rule is that bee colonies shrink in the winter, some less, others more.

That "more" does not occur if the brood-free period lasts as long as possible. The earlier the colonies stop brood rearing in the autumn and the later they start again at the end of the winter, the better the colonies start in the spring.

The beekeeper has no influence on the weather; it must be taken as it comes. But the beekeeper does have control over the other factors. The foundations for optimal wintering and therefore for smooth spring development are already laid in the late summer. Inadequate late summer care often only becomes noticeable during the wintering and then the guessing about the causes begins.

Winter consumption of the honey bees

The feed consumption in the winter (in the winter half year) depends on how strong the colony is, when it stops breeding in the autumn and when it starts breeding again at the end of the winter, of course on the weather and of course on the supply of feed. Only as much can be consumed as is also available.

Strong colonies need more feed than weak colonies and are therefore the first to run out of feed if they have been wintered with (too) little supply. They are usually also the lightest colonies when checked in the spring.

The following values serve as a guideline for feed consumption. In September and October (after feeding) a colony becomes 4 to 8 kg lighter. In November, December and January a colony needs about 1 kg of feed each month, when breeding the feed consumption is 1 to 2 kg and in March and April already 3-5 kg per month. That makes between 14 and 23 kg of feed. If the weather permits, a colony can at least partially cover its food needs in March and April with the nectar it has brought in. But one should not rely on this. It is better if the colonies still have enough stocks in the combs in the spring that are left over from the winter stock. Then they can develop quickly and largely independently of the nectar flow and the weather.

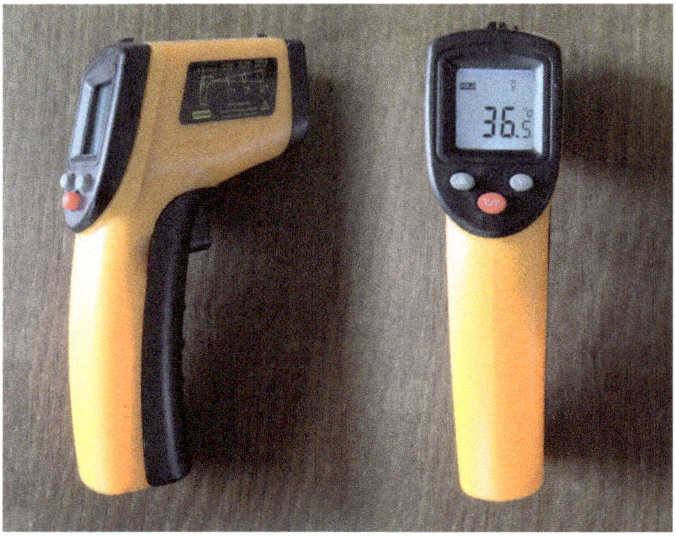

By using a simple digital thermometer you can measure the temperature in the seams where the bee cluster is located and the rest of the space in the hive. Conclusion: honey bees heat themselves and not the hive.

It has no use to keep bees warm in winter

Keeping bees warm used to be a very important issue, also in the bee science of the time (!). In the 19th century, colonies were even placed in dark rooms, cellars or pits during the winter (from November to March) to protect them from the hardships of the winter. At that time, overwintering was considered the biggest problem in beekeeping. This seems to be the case again today, as the frequent losses of colonies in the winter in the 21st century make clear.

Even today, the opinion is widely held in beekeeping that colonies should be kept warm in the winter and therefore should also be close together. For feeding, they are then brought into the "winter seat"; in this, too little attention is often paid to the fact that in a hive there not only needs to be room for the bees, but also and even more room is needed for the food.

A population of 10,000 bees is already relatively light in 9 comb seams in August and needs at least one box of space. In the winter, when there is frost, the bees have packed together in 4 or 5 comb seams and fill the seams at most halfway. The bees only heat the cluster. Little or no heat is lost. It is almost as cold on the unoccupied feed frames as outside. Whoever removes them to keep the colonies narrower and warmer makes a mistake and the wintering becomes a problem.

Widely opened entrances (equipped with mouse wire) ensure that the bees adapt their behaviour better to the outside temperature.

They do not start brooding so early if they not only notice the warmth of the midday sun, but also the cold at night. And this works even better if the colonies are close to the ground or are overwintering on low trestles placed above open wire mesh floors (boxes). Then the wall thickness of the hive or the laborious insulation of hives no longer play a role. The thin-walled wooden hive is sufficient.

When developing the method "vaporising the colonies with oxalic acid from above", it was also observed for ten years (and on several apiaries and under different circumstances and climate conditions) how the colonies overwinter with an empty box placed on top under an inner lid and metal roof. Comparing with "warm head" (= without empty box) shows that the colonies are not harmed if they are kept "cold from below and above" from October to January.

Each treatment includes checking the mite fall. The block treatment with oxalic acid vapour from above in late autumn or early winter is only completed when the mite fall has dropped to the desired level after the last treatment. This can be evaluated after a few days or even 2-3 weeks after the treatment.

Then the consumables can be removed from the empty box and cleaned with water (don't forget to wear gloves). It does not harm the colonies if the empty boxes remain on the colonies for a longer period.

Taking care of young colonies at the end of the year

Young colonies can only grow if they have sufficient food at all times. They take care of the pollen supply themselves. The nectar supply is often so limited in the summer that the young colonies consume more than they can find in nature. The young colonies therefore usually have to be fed continuously. Apart from feeding and timely adding foundations, the young colonies do not require any further care in the summer. In late summer (August/September) they are fed and treated against varroa just like the production colonies.

In principle, the entrance for splits and young colonies must be kept small. Feeding only takes place when the queen has started laying eggs. This does not need to be checked continuously, but (as a rule) only once: observe the development time until sexual maturity, wait for the right weather for the mating flight and give the queen time to provide a large number of cells with eggs. Typically it takes about 30 days "from egg to egg". If you wait 5 weeks, taking out one frame is sufficient. If brood is available, they are fed with sugar water or feed dough. The feed dough can be placed as a flat slab on top of the frames and covered with foil and the concave side of the inner lid. After two to three weeks the young colony is checked and food dough is placed again.

Sugar water (or syrup) can be given to the growing young colony in flat plastic dishes that are placed on the mesh bottom box against the occupied frames (the use of Tetrapaks as in former days has now been abandoned, because there are always colonies that gnaw a hole at the bottom of the package. A flat plastic dish on the mesh bottom box with syrup or sugar water and a handful of "vegetables" - so that the bees do not drown - works much better).

Don't forget swimming aid in/on the sugar water! More elegant than using flat dishes, although more expensive, are the closed feeding frames that are one or two frames wide and can be filled with feed dough, honey, syrup or sugar water for feeding.

Don't forget swimming and climbing aids either!

Diseases of the honey bee

We know of a large number of diseases that occur in honey bees. There are those that only affect the brood and others that only affect adult bees. The pathogen of by far the most important disease, the varroa mite, that parasitizes both adult honey bees and bee brood. However, varroosis is primarily a brood disease. The mite can only multiply in the brood and does the most damage there.

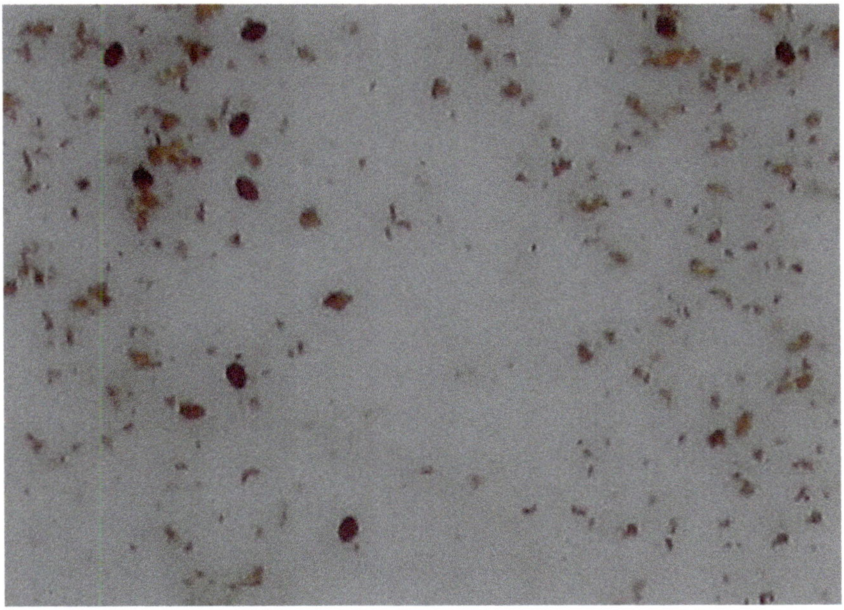

A natural varroa mite fall on the inspection board

The bee colony has naturally developed in such a way that it can control almost all diseases. An exception is varroosis. This disease will be discussed in more detail in a separate chapter.

In the bee colony, thousands of bees live close together. If a few bees become ill, the risk of infection is very high, also because the place where the bees stay is at the same time the storage place for the supplies and the place where the brood is cared for. The supplies are stored close to the brood nest at temperatures above 30°C. but even then they do not spoil! The conditions in the bee colony should actually provide ample space for micro-organisms, fungi and bacteria to grow and multiply.

The fact that bee colonies can nevertheless remain healthy under these circumstances makes it clear that they generally have the risk of infection under control. Unfortunately, not always. Colonies can become ill and even so badly that they suffer from the disease and can even perish. Sick and dead colonies have a high infection pressure. Pathogens can be transmitted to other colonies through evaporation or robbery. This requires the beekeeper to prevent evaporation and robbery as much as possible. Dead colonies must be cleared before they are discovered by neighbouring colonies and are cleared out. This applies in particular to the colonies that have usually perished in late autumn and winter due to inadequate varroa control treatment. On the continent (Wst-Europe) this amounts to approximately 15% of the colonies kept by beekeepers in long-term averages.

The causes of most diseases are present in every bee colony. The outbreak of many diseases is favoured by the following factors:

- inadequate comb hygiene
- lack of food (hunger) and
- unfavourable weather conditions for the bees.

As far as the weather is concerned, the beekeeper cannot do anything. The beekeeper and the bee colonies must accept it as it is. There is usually nothing to fear when the weather plays tricks in the winter or spring. If the food supplies and the colony strengths are in order, the bee colonies can cope with the vagaries of the weather.

When problems occur, it is often a matter of comb hygiene. Pathogens can easily nest in the brood combs, which is why they should only remain in the colony for a limited time. Old dark combs have no place in the colony. They must be removed when the colonies are reduced in late summer. This involves some effort. If you only remove the honey super when reducing the size, it will be considerably easier.

Who would not be tempted? The annual renewal of the combs in the brood space is nevertheless or therefore urgently recommended; because prevention is better than cure! And again, with the "right hive" and the "right beekeeping method", changing them is easier.

Prophylaxis also includes ensuring that the colonies never run out of food.

The workers produce saliva with antibiotic-active substances. They use this saliva multifunctionally:

- When cleaning the cells, the walls are impregnated with it.
- Nectar and honeydew are enriched with their own antibiotic-active substances when thickening. It is precisely this addition of saliva that makes the honey valuable.
- Antibiotic-soaking substances are also found in the feed juice and ensure healthy growth of the larvae.
- The sensitive pollen is preserved by salivation so that it does not spoil even when stored at warmer temperatures.

The main diseases and pests of the honey bee in brief

designation	Vairimorha (Nosemosis)	Acaraiosis	Varroosis	Chalk brood	Malignant foul brood	Wax moths
or	Spring slump or spring disappearance disease	Respiratory mite disease			American foul brood	
causative agent	Vairimorpha apis V. ceranae	Trachea mite	Varroa mite	Acospaera apis	Bacillus larvae	Large and small wax moth
happens to	adult bees	adult bees	brood and bees	brood	brood	especially brooded combs
also survives outside the colony	yes	no	no	yes	yes	yes
symptoms of disease	colonies become weaker	crawlers and hoppers	natural mite fall	mummies	dead brood	feeding traces, spider webbing
defense forces	available	available	not available	available	available	avail-able
the course of the disease is promoted by						
bee combs turning:	inhibited	inhibited	promoted			

The workers produce saliva with antibiotic-active substances. They use this saliva multifunctionally:

- When cleaning the cells, the walls are impregnated with it.
- Nectar and honeydew are enriched with their own antibiotic-active substances when thickening. It is precisely this addition of saliva that makes the honey valuable.
- Antibiotic-soaking substances are also found in the feed juice and ensure healthy growth of the larvae.
- The sensitive pollen is preserved by salivation so that it does not spoil even when stored at warmer temperatures.

If saliva is insufficient, cleaning behaviour helps further. This is another general weapon of the honey bee colony that not only concerns keeping the bee house clean but is also used to remove sick larvae and pupae, and in time before the pathogen can spread. Of course, this does not always work. Then the cleaning behaviour can have a negative effect. The cleaning bees can infect themselves when removing dead brood and dead bees as well as when licking the droppings of sick bees, become sick themselves and pass on the disease.

Even if this occurs, nature has provided measures. Sick and old honey-bees usually leave the hive of their own accord and then perish outside. This behaviour greatly reduces the risk of infection and is particularly evident during the brooding period. In sick colonies, a high bee turnover can lead to self-healing.

What promotes and slows down the outbreak of diseases in the honey bee colony?

disease-inhibiting properties	disease-promoting properties
▸ antibiotic substances in bee saliva (and therefore also in honey, feed juice, propolis and pollen)	▸ thousands of bees live together side by side
	▸ bee shelter, supplies and brood care in one place
▸ grooming behaviour reduces risk of infection	▸ cleaning behaviour promotes infection
▸ adequate food supply	▸ volatilisation and robbery

▸ faecal bladder	▸ causative agents are probably present in every population or almost in every population
▸ bee combs turning	▸ brood combs are nesting places for pathogens

By far the most dangerous disease of honey bees is varroosis, which is caused by the mite Varroa destructor. A honey bee colony is not sick if it has varroa mites. The disease only breaks out when the number of mites in the colony has reached a high infection rate of many thousands. The honey bee colony cannot defend itself against the varroa mite. With regard to varroosis, it is dependent on the help of the beekeeper from year to year.

Other diseases such as Vairimorpha (before 2020: Nosemosis) and Acaraiosis have spread fear and trembling in former times until beekeepers have learned to care for the colonies in such a way that they can no longer cause great damage. In addition to varroosis, the malicious foulbrood plays an important role.

The malicious foulbrood

The vicious foulbrood - it is also called American foulbrood, but does not originate from America at all....- is a disease for which notification is mandatory. (In some countries clinical infection with American foul-brood must be reported to Health Security Authorities or to a vet or a veterinary organisation). When the disease occurs, exclusion zones are set up in the vicinity of the infected apiaries (quarantine) from which no infected colonies may be dragged away in order to prevent the spread of the disease. Such a exclusion zone is only lifted when the disease has disappeared. The affected apiaries are officially remediated.

In overseas areas the disease is widespread, which means that honey from such countries often contains traces of the pathogen. There is a great risk of infection when feeding such honey. Therefore, honey from foreign sources should not be fed to honey bees.

The pathogen of the malicious foulbrood (American foulbrood) is called Paenibacillus larvae. The larvae are infected by the nurse bees when they make their feeding juice from honey that has been contaminated with spores. The nurse bees can also ingest the spores when cleaning the cells if larvae have previously died from the disease.

In addition to the malicious foulbrood, there is also the benign foulbrood (benign[]) which is also called the European foulbrood or also just sourbrood because the infected larvae smell sour. This stench is not suitable for a certain diagnosis. In the event of dead brood, one should always inform the expert on duty of the beekeepers' association or another expert in charge.

Vairimorpha (before 2020: Nosemosis)

Vairimorpha probably occurs every spring. The disease usually remains unnoticed by the beekeeper. Vairimorpha is an intestinal disease of adult bees and is caused by the sporidae Vairimorpha (Nosemosis) apis. In the past few decades, the Asian relative Vairimorpha (Nosemosis) ceranae that originally lived on the Eastern honey bee has also spread in Europe and has probably displaced the native Vairimorpha (Nosemosis) apis. In contrast to the very first assessment, Vairimorpha ceranae does not seem to be more dangerous than Vairimorpha apis.

The cause(s) of Vairimorpha are present in every bee colony. The infestation usually increases in the spring, reaches its peak in April/May and then drops to almost zero in the fall. In the case of a very strong and persistent infestation, many honey bees die prematurely before they are replaced by young bees. The result is that the colonies become weaker in the spring instead of growing. That is why Vairimorpha is also called "spring sluggishness" or "spring disappearance disease". In the case of a mild course of the disease, the colonies do not need to be treated for Vairimorpha, because Vairimorpha is a factor disease that only becomes a problem in unfavorable weather and site conditions.

Even then, self-healing is possible in the affected colonies.

The causative agent establishes itself in the intestines of adult bees. Its multiplication is promoted by high temperatures and protein conversion in the intestines. That is why mainly nurse bees - they stay in the warm brood nest and produce protein-rich feed juice - are affected. The infected bees defecate frequently. In persistent cold weather, many do so in the hive.

The excrement contains Vairimorpha spores. The hive bees become infected when they lick this excrement and become ill when they subsequently become active as nurse bees. A high bee turnover protects against an outbreak of the disease and reduces the infection rate of the bee colony.

Vairimorpha, then stil called Nosemosis, was considered very dangerous until the

1970s. For treatment, often also for prevention, the antibiotic Fumidil was used in the past, which was taken off the market over forty years ago. Nevertheless, no epidemics broke out as described earlier. Perhaps this is because beekeeping has changed over time and a fundamental change in the care for the colonies has also ensured that honey bees are less susceptible to spring slump or spring disappearance. In the 1960s and 1970s, the management practices with the change from "back treatment" to "top treatment" triumphed, which also resulted in a higher bee turnover in the honey bee colonies. In addition, with the spread of the varroa mite in the 1980s, the importance of other diseases was downplayed.

Acaraiosis

Honey bees, like other insects, have a finely branched system of airways, the so-called tracheae, which supply all organs with oxygen. The small mite Acarapis woodi (Acari = mites, apis = bee) lives in the thicker airways in the honey bee chest, which is also called the trachea mite because of its habitat. The mites feed on bee blood that they obtain by piercing the wall of the airways. The females lay eggs from which larvae emerge that are adult and mate after 2-3 weeks. If the airways become too narrow, the mites move to other honey bees. The infected bees have breathing problems and are therefore unable to fly. They are noticeable in front of the entrance as crawlers and hoppers.

A strong bee turnover leads to self-healing because the infected honey bees leave before the mite offspring mature and can move on to other bees.

Chalkbrood

The chalkbrood disease of the Western honey bee Apis mellifera is a mycosis that exclusively infects the honey bee brood. The causative agent is the fungus Ascosphaera apis. The fungus spreads by spores. The maggots are infected when they are fed with food containing spores. The spores enter the midgut of the maggot and germinate there. The fungus initially develops in the digestive system. Later, the fungus pierces the maggot with its mycelium (fungal twig) and penetrates the body wall to the outside. The sick larva develops into the stage of the stretch maggot or prepupa, then it dies. The dead maggots are covered with a white fungus. On the surface of the hard maggot mummy, sporangia develop when male and female fungal mycelia come together. They colour the mummy grey-black or grey-green. When they are ripe, they burst open and release an unashamedly large number of spores.

Chalkbrood is considered a factor disease. Weak colonies are considered particularly susceptible. Its occurrence is favoured by cold and humidity. In humid years the disease will appear stronger than in drier years. Contrary to this assessment is the observation that chalkbrood sometimes causes greater problems in countries with a semi-arid climate such as Greece, Italy and Spain than in countries with a humid climate. The course of the disease is often regarded as relatively harmless and the extent of damage in most cases is considered to be minor. It is generally believed that with good beekeeping and colony care chalkbrood can be kept under control and eliminated.

In the event of a strong occurrence of chalkbrood, the following measures are recommended:

- remove the infected brood frames,
- place the colonies on new combs,
- provide the colonies with a different queen,
- place the colonies in a more favourable location,
- stimulate the cleaning behaviour of the colonies by feeding during periods when there is no flow.

The most effective method is to exchange the queens of the colonies with chalkbrood.

Varroosis

The instigator of varroosis is the mite Varroa destructor. This mite is a serious problem for beekeeping worldwide, especially in temperate climate zones. There it is generally the case that honey bee colonies of the western honey bee Apis mellifera that are suffering from varroosis die if they are not treated. Only in the tropics or subtropics can bee colonies of the western honey bee Apis mellifera (also) be kept without treatment against varroosis despite varroosis contamination.

In the search for the cause of this natural varroosis resistance or varroosis tolerance, several hypotheses and theories have been developed and based on this, the selection of varroosis-resistant or/and varroosis-tolerant honey bee colonies in the northern hemisphere of the world has been attempted.

Meanwhile, there also seem to be honey bee colonies in temperate climate zones that survive without treatment against varroosis, such as in the USA, but also in Europe, such as in Norway, Sweden, Finland, Wales, the Netherlands and southern France. The reports of success have also led to the testing of offspring queens from the supposedly resistant colonies elsewhere, so far with the same result. The colonies that were bred with these queens were unfortunately neither varroosis-tolerant nor varroosis-resistant.

Consequently, the ability of the colonies on site to survive without varroosis treatment is not (only) in the genes of the queen. This must be kept in mind when looking for a beekeeping without varroosis treatment. There are many ideas. A great many of them can be found on the internet. Dr. G. Liebig is also concerned with these ideas - as on the following pages of this book - and is also looking for new ideas on a solution of the problem. This goes hand in hand with the aim of optimising the ever-necessary treatment against varroosis. In both cases, attention must be paid to the biology of the varroa mite and its host.

About the life cycle of the varroa mite

In temperate climate zones, the losses of colonies that occur mainly in winter can be attributed primarily to inadequate treatment against varroa in late summer/autumn. The eastern honey bee Apis cerana, the original host of the varroa mite (and of Nosemosis ceranae) is varroa-resistant. In Apis cerana, the varroa mite reproduces exclusively in drone brood. The mite remains sterile in the worker brood, so that very few mites (at most a few hundred) are present in the colonies of Apis cerana.

In Apis mellifera colonies, the varroa mite multiplies in both drone brood cells and worker brood cells. During a brooding period - from March to October - a varroosis infestation of more than ten thousand mites can build up, which inevitably results in the outbreak of disease if the beekeeper does not intervene. The treatment of the colonies against varroosis is necessary every year in late summer and autumn, before the varroosis infestation exceeds the damage peak, which is considerably lower in autumn than in summer, which can easily be deduced from the colony development of the honey bee as is common in temperate climate zones.

In the autumn the "winter bees" are brooded. They live longer than the "summer bees" and after the wintering in the spring they raise the first generation of "summer bees" for the coming year. "Winter bees" can only accomplish this feat if they survive the winter. This is now guaranteed if they were born healthy in the autumn and have been protected from varroosis contamination and its consequences; because "winter bees" that have been contaminated with varroosis in the brooding stage die prematurely. If it has happened to too many "winter bees", a good wintering of the colony is at stake.

In the summer, the then still strong colonies tolerate more mites than in the autumn. The degree of varroa infestation should not exceed 10%. More than 90% of the "winter bees" must grow up without varroosis infestation.

Reproduction and infestation development

The varroa mites suck on the bee brood and on the adult bees. The reproduction of the females takes place exclusively in sealed (capped) brood.

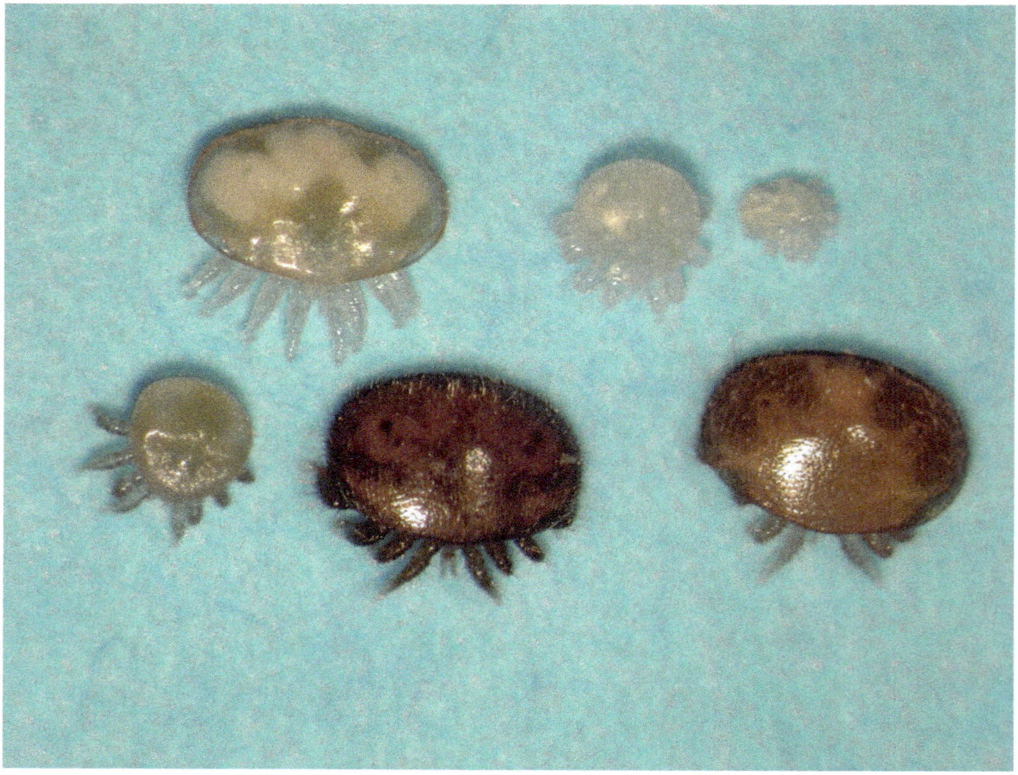

The mite brood from a cell where the bee that grew up in it is about to emerge. The dark brown mother mite among her offspring. Only the eldest daughter (right) is viable, all others fall prey to the cleaning bees.

Photo Dr. G. Liebig

The fertilised female slips into a brood cell shortly before it is sealed and starts laying eggs about two days after the brood cell is sealed. The rule of thumb is: "one egg every day".

A male develops from the first egg, and females develop from the following eggs. Males and females need about six to nine days to reach adulthood. Before the infected honeybee emerges, the adult female offspring are fertilised by the male or the males.

If only one mother mite has entered the cell, close family mating takes place. This is the rule for moderate varroosis infestation. This continuous incest should actually have a negative effect on the vitality of the mite(s). But that does not seem to be significant, because the virulence of the varroa mite has remained unchanged to this day.

Only those daughters are viable and can reproduce which are adults and have been fertilised before the infected honey bee emerges. The fertilised females first spend a few days on the young hive bees before they crawl for reproduction into bee brood cells that are going to be sealed. They do not have to go looking for such a cell or walk to it, but simply wait until the honey bee on which they have grown up and with which they emerged takes on her task as care bee and in this function takes part in feeding older bee larvae. Then, on a suitable occasion, the mites only have to step into a brood cell that is soon to be sealed.

After emerging, mites can also change bees. This is certainly true for mites that have grown up in drone brood. The mites that have come "into the world" with or on a drone probably use the or a feeding of their host for a "host change". The duration of the stay of the mites on the bees - the phoretic phase - depends primarily on the brood size of the colony. In strongly brooding colonies or in the months with a lot of brood, this period is shorter than in weakly brooding colonies or in the months with little brood.

The mother mites can reproduce twice, some even three times. The number of offspring depends on the sealing time of the brood, which lasts about three days longer in drone brood than in worker brood. In the drone brood, adult offspring can therefore also develop from the eggs laid later, so that two to four fertilised daughter mites emerge with the mother mite. In the case of worker brood infestation, there is or are one or two.

Since drone brood is also clearly more heavily infested with mites than worker brood, the number of mites in the honey bee colony increases particularly rapidly when drone brood is available.

The following conclusions can be drawn from the development cycle of the mite and its reproduction behaviour:

- During the brooding period, the number of mites in a colony can double or triple monthly. If there are no brood breaks between March and October that interrupt the mites' reproduction activity, the number of mites can increase by more than a factor of 100 during the vegetation period. From a mite that overwinters in the colony, an infestation of at least 100 mites develops into by the autumn, from 10 mites at least 1,000, from 100 mites more than 10,000 mites, if the colony does not collapse under its mite burden before then.
- The development of the contaminattion is closely related to the development of the colony. The more brood a honey bee colony rears from spring to autumn, the faster its mite count increases. However, this also means that if the breeding activity starts four weeks earlier due to weather conditions and location, there will be approximately (or more than) twice as many mites in the colonies in autumn than otherwise, or the limit of the critical infestation strength will be exceeded a month earlier.
- If the number of brood cells and bees decreases in late summer or autumn, this results in an explosive increase in the relative degree of infestation of brood and bees. Honey bee colonies that already have an (at that time still not very harmful) infestation of approximately 100 mites in the spring and that develop strongly in the spring and summer so that the varroa mites can also multiply well, are already in great danger in August or September.
- As long as brood is available, there are mites in the sealed/capped brood and on the adult bees; the ratio depends on the brood size. In the summer, when the colonies have more brood cells than bees, approximately three quarters of the mites are in the sealed brood. This applies to both old and young colonies to an equal extent.
- In addition to the offspring that can reproduce, there are also offspring that have not yet completed their development by the time the infected young honey bee emerges or that have not been fertilised and are therefore not viable. These offspring fall down when they emerge and, together with the mother mites that die of old age, cause the "natural mite fall". The more mites reproduce in the brood, the higher the natural mite fall (and vice versa).
- Varroosis is primarily a brood disease. Workers are consistently damaged if they were parasitised in the pupal stage. Their lifespan is shortened. They die earlier, which is particularly evident in winter bees. Even if they are freed from the mites

immediately after they emerge, they remain damaged. Therefore, excessive parasitisation of the winter bee brood must be avoided. In late summer and autumn, constant attention must be paid to ensure that the varroa infestation does not exceed the damage threshold and that timely action is taken.

Characteristics and rules of varroosis infestation development in brief

The infestation at the end of the year is determined by the following characteristics:

- by the initial infestation in the spring:
- This is the number of mites in a colony in the spring when it starts brooding and the overwintered mites start to multiply. It is impossible to determine the initial contamination exactly, unless the colony is killed. The mite fall caused by the treatment(s) at the beginning of the winter of the previous year and/or the natural mite fall in the winter provide an orientation. The conversion factor for this is 500; i.e. if in the winter in a non-breeding honey bee colony without treatment a mite falls daily, then there are approximately 500 mites in the colony. When diagnosing the fall in the winter, care must be taken that the mite fall is not disturbed by the fall of dead bees.
- This should also be taken into account when mites fall due to treatment(s). The mites that land on the falling dead bees only fall through the bottom grid when the dead bees are carried away or cleared away. If in January/February not a single mite falls on the inspection board for weeks without treatment, the colony has overwintered with very few mites - no more than 10 or 20. Such a low initial contamination can be achieved by repeated evaporation of oxalic acid in late autumn/beginning of winter without damaging the colonies, because the evaporation of oxalic acid is tolerated very well by the bees. On the other hand, the treatment of honey bee colonies with an oxalic acid solution by spraying or dripping is less tolerated by the bees. The following applies to spraying or dripping: "twice is once too much".
- by the fertility of the mites during the brooding period:
  - proportion of mite females that can reproduce,
  - number of offspring per mite female,
  - number of reproduction cycles per mite female.
- These quantities (characteristics) also depend on the population development, in particular on the duration of the brooding period, on the size of the brood nest and the occurrence of drone brood.
- by the mortality of the mites:

- natural mite fall.
- by mites "export":
- Removal of contaminated brood or contaminated honey bees, loss or escape of contaminated bees and drones, killing of mites by treatments after the honey harvest.
- and by mites "import":
  • Flying in of contaminated foreign honey bees and drones, introduction of mites by the colony's own bees that rob [steal honey] from other colonies.

The infestation development of a colony is therefore not only determined by its internal but also by external factors that are independent of the colony. In connection with this, the search for genetically anchored characteristics of honey bee colonies that determine the receptivity or sensitivity of honey bee colonies to or for the varroa mite is difficult. Worldwide, also in Europe, intensive searches are being made for honey bee colonies that are distinguished by less receptivity (resistance) or less sensitivity (tolerance) to the varroa mite. "Success reports" are not lacking...

Initially, such reports were limited to beekeeping in the tropics and subtropics, later individual cases from temperate climate zones were also added. Many of these reports can be found on the internet.

The manner, origin, causes and order of magnitude of this possible resistance or tolerance are not yet clear. The discoverers often emphasize that only ecologically adapted bees can develop varroa resistance and tolerance and only then if these colonies are kept isolated and not treated. Those who blindly follow such recommendations and trade instructions usually suffer shipwreck; because so far the "success reports" have always turned out to be misinterpretations in critical research. Examples are the hypes

- - regarding the "primorski bee", which was promoted as a resistant origin and from which not only queens but also breeding material were sold expensively for years,

- - regarding the "killer factor", which was even mentioned as a breeding characteristic in the breeding guidelines of the National Beekeepers' Associations, although it was known from the outset that "the number of damaged mites in the fall" is not an indication of varroa resistance.

- - regarding the "small cells", in which the varroa supposedly cannot multiply. Appropriately stamped middle walls were and are offered in the trade. At the beginning of the "varroa era", there were plastic combs with enlarged worker brood cells on the market, in which supposedly no varroa multiplication would take place. Research showed that both assumptions are false.

A similar fate also befalls the inventions that were claimed to be able to control the varroa mite without the use of chemicals. Examples are "Heinrich's intermediate bottom", the "Muller plank" and "VarroaCatch". In their application, the sealed brood is separated from the open brood by a bee- and mite-proof mesh. The mites that emerge with the brood are attracted - in the inventors' concept - by the smell of the open brood on the other side of the mesh and are caught in the mite-proof mesh after passing (now suddenly on foot...) through the sealed mesh. The trap "works" particularly well if the combs with hatchable brood are placed above the combs with open brood (and with the queen). Of course, only the natural mite fall that occurs with the brood that emerges lands in the trap. Some of these mites are still alive, but they are not viable.

Another example is the "rotating hive", an invention from Hungary. In the rotating hive, the brood space under the honey super is constantly rotated - by hand or with a motor - which should prevent the varroa mites from multiplying in the brood. It is said that the offspring with their thin skins are crushed to death by the pupae in the narrow cells when rotating. Rotating is also said to prevent swarming, which is also a fact and is easy to carry out in experiments. If open swarm cells are turned upside down, the queen larvae drown in their food juice; because they can't swim, so to speak. Nevertheless, it is not worth changing the hives or the operating method to "salto beekeeping".

The "tilt control and (if necessary) breaking the swarm cells" concept is less effort to prevent the departure of swarms.

For combating the varroa mite with ultrasound, the "ultrasonic-o-mat" was on the market in the nineties and thirty years later the "Varroa-Killer-Sound". In tests, both devices showed no effectiveness.

The heat treatment has also not been able to prove itself as effective. The result is very small, the effort is great. In addition, negative side effects can occur. For example, heating also causes the stored food to drip and thus triggers robbery.

What the damage looks like

In an individual honey bee, varroa infestation leads to a clear loss of weight, to a shortened lifespan and to a failure of performance. Infected honey bees are restless, their social behaviour is disturbed.

With the degree of infestation of the honey bee colony, the multiple parasitism of the brood cells (the intrusion of two or more females) also increases, which can occur particularly often in late summer and autumn when the brood nest shrinks due to increasing varroa infestation. As a result, the number and proportion of clearly and visibly damaged bees (mutilated wings, mutilated abdomen) increase rapidly.

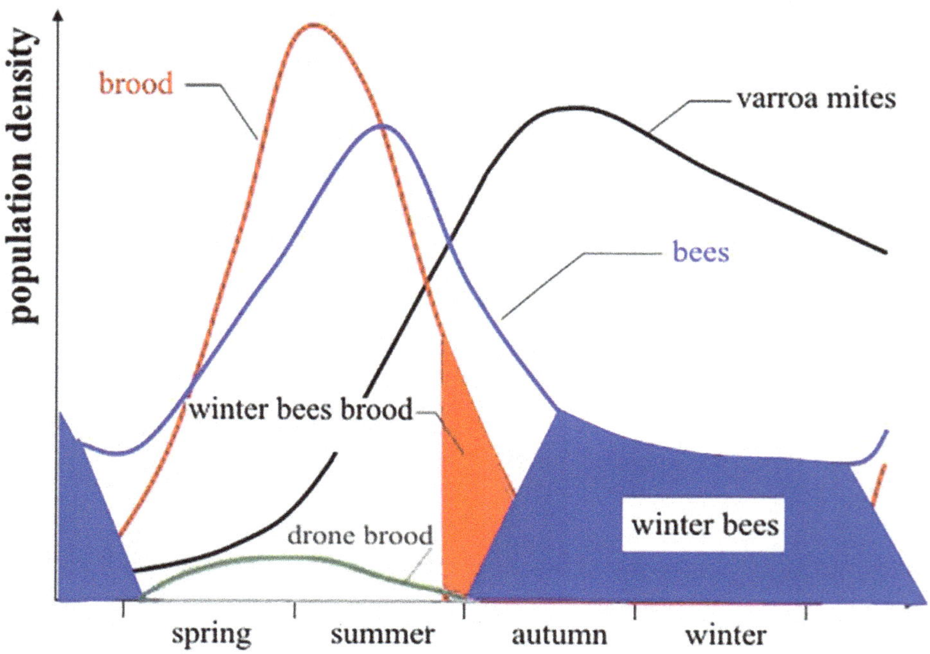

Comparison of the development of the honey bee colony with the infestation development during the year (schematically depicted). The critical period for the honey bee colony is in the autumn. In July/August/September the brood curve decreases, the honey bee colony decreases in size, the number of mites increases rapidly. The result: the relative infestation rate of the honey bees increases like a thunderbolt and that precisely in the time that the valuable winter bees are developed.

(Picture after Dr. G.Liebig)

In the case of severe infestation, the care for the brood is also neglected. As a result, other brood diseases (secondary diseases) can occur. At the same time, many sick bees die, the honey bee colony becomes weaker by the day. It collapses. Often the collapse occurs in the autumn (September, October, November), during or after feeding with syrup or sugar water. In the collapsing colony, robbing bees have easy access and take the food. When passing the brood nest, the mites jump onto the robbers who then carry the mites home together with the stolen food. Within a few days, the mite population in a colony that robs a colony damaged by varroa can increase by several hundred mites, which negates the success of an earlier late summer treatment.

This external infestation pressure is particularly high during the period of poor forage during and after winter feeding and has become a "re-invasion". For this reason, a synchronous varroa treatment of all colonies in an association area is often recommended and here and there also required by the government so that a "re-invasion" does not occur.

However, this concept is nonsense as such. The extent of mite introduction is extremely overestimated. Repeated checks of this process have shown that less than 5% of the mites from a colony collapsing due to varroa end up in the neighbouring colonies, and even then only if they are exterminated by them, which can be prevented by keeping the entrance small. On (almost) every apiary it happens (almost) every year that a heavily infected colony stands next to a weakly infected colony without it leading to a mites being introduced.

The difference in varroa infestation between colonies on an apiary and also between different apiaries is not caused by the introduction of varroa mites. A higher varroa infestation in the autumn can rather be traced back to the fact that the treatment previously applied in late summer worked less well than expected. This occurs relatively often, also or especially when the approved products are applied according to the instructions. The "re-invasion" is then used as an excuse.

Varroosis infestation control

Successful beekeeping requires knowledge and consideration of the lifestyle of the honey bee colony and the varroa mite. This includes being able to assess when the parasite becomes a danger to the honey bee colony and before damage occurs. Anyone who sees deformed honey bees crawling around in front of the entrance or mites walking over the combs and brood suffering from varroosis or secondary diseases during a normal colony control and is surprised by this, has simply been asleep. It is often too late for measures to save the colony.

The development of the contamination must be monitored and limited in time so that it does not get to that point. This monitoring is absolutely necessary in late summer and autumn; because in this period the degree of damage is relatively low and can easily and quickly be exceeded.

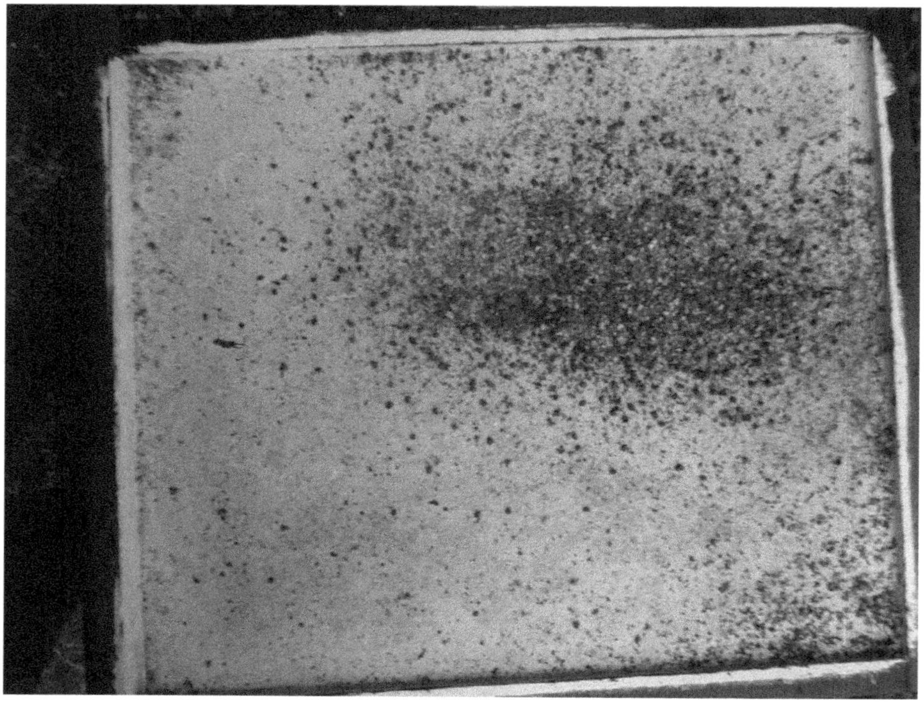

A inspection board. Not so many mites to be seen, at this moment, but then a lot of pollen instead.

The degree of contamination of a honey bee colony can be assessed by a simple diagnosis of the fall or by examining bee and brood samples with great difficulty.

In summer, approximately three quarters of the mites are in the brood. The adult mites - old and potential mother mites - only remain on the adult honey bees for a few days. Shortly before sealing, they move into the brood cells to multiply there. When the young hoeybee emerges, the mother mites and her offspring leave the brood cell with her. The offspring consist of the first-born son and the daughter mites. The oldest daughter mite is fertilised in the brood cell while still in the brood cell, in the case of a single contamination always by her brother. Only the mother mite and her fertilised daughter are viable and can reproduce, the younger and immature daughters and the male are no longer viable. Together with the old mother mites that can reproduce up to three times, they cause the natural mite fall. Based on the mite fall, the varroa infestation of the colony can be estimated. For this, it is necessary to hold the colonies above a gauze base with a mesh width of approximately two millimeters. A inspection board is inserted into the mesh bottom box from behind. The inspection board is slid under the honey bee colony for a period of three to ten days. The shorter the inspection board is under the colony, the easier it is to discover the mites in the fall. This is especially true for the time in spring and early summer when the colonies are very active and therefore a lot of waste falls down.

Guard bees at the entrance of a small entrance. At least four bees are contaminated with varroa mites. The mites on the abdomen or the mites underneath are hardly noticeable. It is therefore not possible to assess the degree of infestation by looking at the bees during a colony check. The laborious plucking of sealed (capped) brood also does not provide a certain diagnosis.

Photo Dr. G. Liebig

Sometimes it happens that the waste is searched for food by ants or earwigs, which can also grab mites and carry them away. The wind blowing through the entrance can also blow the light mites to the edge or into a corner of the inspection board, which can then easily be lost when pulling it out. Therefore, for an accurate count, the inspection board should have a rim all around so that it is also possible to grease the inspection board with rapeseed oil to make it difficult for the ants or earwigs to remove mites from the inspection board. It is usually sufficient to slide the inspection board in for a short time so that the ants and earwigs do not get used to using the mite precipitate as a food source.

If it is very windy when counting mites, it is best to cover the pulled-out inspection board with a second inspection board and find a sheltered spot for counting. You can also stack the inspection boards on top of each other and then carry them into the house or car and make the diagnosis there. Another practice is an alternative without wind and especially then if there are a lot of mites on the inspection board after a treatment: make photographs of the pulled-out inspection board with a digital camera and then count the mites on the screen at home.

The natural mite fall consists of dark brown (adult) mites and lighter mite stages that are usually yellow, yellow-brown to light brown in colour.

Sometimes there are also white stages that can be found more easily if you put on a magnifying glass while counting. The light mites in their fall have not been able to finish their development in the brood cell and are therefore not viable and have fallen down when the honey bees emerge or afterwards when cleaning the brood cells. The dark brown mites are probably old mother mites.

When counting, the mites that have the characteristic broad-oval shape of the varroa mite are counted. One counts the dark and the lighter mites that usually predominate in numbers when the colony being examined has brood. If only or predominantly dark mites fall naturally in the summer, there is reason to suspect that the colony in question has no brood. It must then be checked to see if it has a queen. There is a correlation between the natural mite fall (mites per day) and the varroa infestation of the honey bee colony (number of mites) that is not particularly close, but is sufficient to recognise whether danger is lurking.

The order of magnitude of the conversion factor depends on whether brood is available or not. In addition, the degree of contamination plays a role.

For the period from May to September, the conversion factor is in the order of magnitude of 100 to 300. The higher the natural mite fall, the smaller the conversion factor. 1 mite per day means 300 mites in the colony rather than 200 or 100. If 100 mites fall daily without treatment, there are not 30,000, but "only" 8,000 to 15,000 mites in the bee colony.

In collapsing colonies, the conversion factor is even lower. If the colony is doing badly, the mites are also doing badly and their mortality is increased. Colonies that collapse due to varroosis can lose more than 1,000 dark (!) mites every day in their "final phase".

In late autumn or in winter when the colonies have no or very little brood, the conversion factor is in the order of 500.

The damage thresholds

Our honey bee colonies are sensitive to the varroa mite and therefore they have to be treated regularly. They become ill if the varroa infestation exceeds the damage threshold and they die if they are not treated in time after exceeding the damage threshold. These statements will remain generally valid for the time being.

But where is this damage threshold and when is it reached? The opinions of beekeepers and experts differ on this, and also because the damage threshold has no fixed size and depends on the population strength (number of bees and size of the brood), but also on the time of year. In the summer, the colonies tolerate more mites than in the autumn and winter. These three damage thresholds can be linked to the natural mite fall.

In summer, the natural mite fall may not exceed the value of "100 mites per day". The damage threshold in autumn is "10 mites per day" and in winter only "1 mite per day". If these damage thresholds are exceeded, this does not automatically result in the death of the honey bee colony in question if it is subsequently freed from its mite burden by a carefully carried out treatment. It will only enter or leave the winter weaker. f the damage threshold is clearly exceeded, it will be so adversely affected that it can perish in the winter.

The damage thresholds described can also be expressed in absolute mite numbers, taking into account the population size. For example, a colony that has 25,000 honey bees and 10,000 brood cells at the end of August will not be too adversely affected by a varroa infestation of 15,000 mites if it is extensively freed from this infestation in August and thus before the winter honey bees are raised. This can be achieved with a one-time treatment with formic acid if it has a correspondingly high effect.

Each colony should enter the winter with (clearly) less than 1,000 mites. With a wintering strength of 10,000 bees, the relative infection rate would be below 10%. However, a varroa infection of 15,000 or more mites in August does not come out of nowhere. It builds up during the spring and summer, which results in honey bee colonies suffering from a heavy varroa burden in the summer before the period in

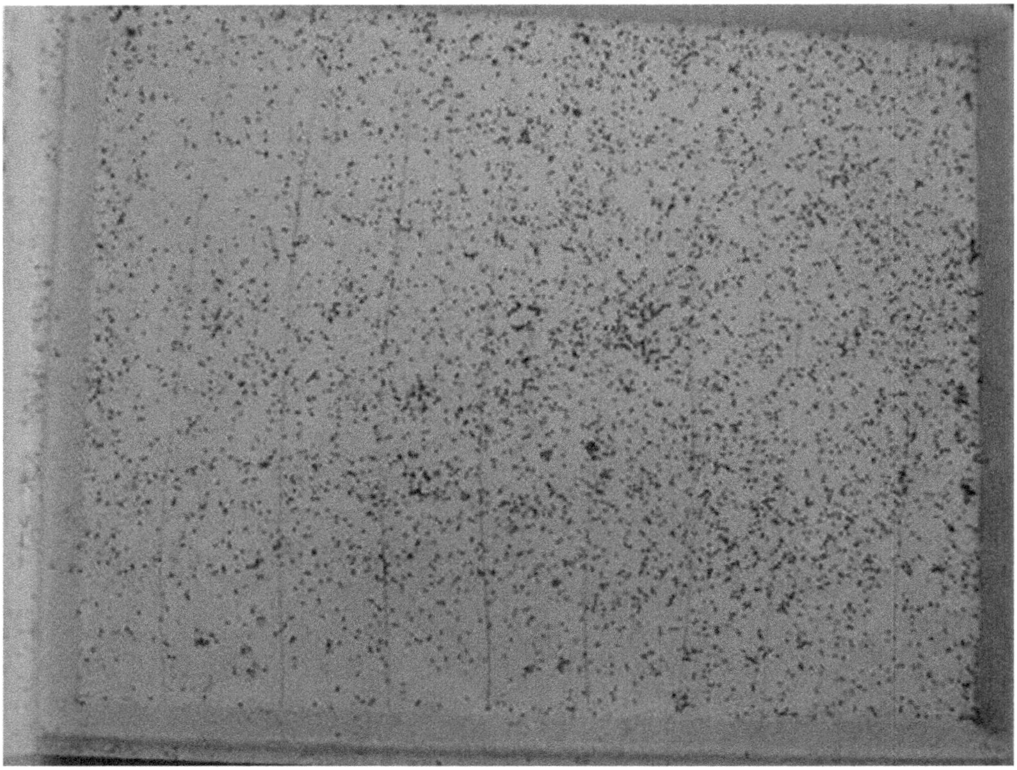

Mite fall after treating a brood-free "brood colony" with oxalic acid. The entrance reduced with a strip of foam rubber was in the photo on the bottom right. More mites fell near the entrance than further away, a total of 3,424.

Photo Dr. G. Liebig

which the winter bees are brooded, which causes their premature collapse. In such cases, an exact measurement of the damage threshold is made impossible afterwards because the collapsing colonies lose not only their bees but also a large part of their mites that have caused the damage, which also includes the occurrence of secondary brood diseases. If the investigation only starts at the stage of collapse, the damage threshold is easily underestimated and the influence of secondary diseases is overestimated.

The large difference between the damage thresholds in summer ("100 mites/day") and autumn ("10 mites/day") reveals why it happens time and again that honey bee colonies nevertheless succumb to varroosis, despite supposedly effective treatment in the summer when "many mites" have fallen.

In the summer treatment, it is not important how many mites are killed, but how many mites remain in the colony; because the mites that survive the summer treatment multiply in the winter bee brood that is reared from the end of August/beginning of September and thus double or even triple their number every month.

After a reduction in the natural mite fall due to the summer treatment from 50 mites per day to 5 mites per day (which corresponds to an effectiveness of 90%; this value is rarely achieved with a treatment in the summer), the numbers can be as high as ten mites per day in September due to the multiplication of the "residual mites" and even 20 mites per day in October, which means a clear exceedance of the damage threshold in the autumn. Therefore, in addition to the summer treatment in August, an autumn treatment in September is also necessary. Both treatments must "sit".

The summer treatment should ensure that the varroa infestation is clearly pushed below the autumn damage threshold. The autumn treatment aims to ensure that the multiplication of surviving mites does not lead to exceeding this damage threshold and that the colonies enter the winter with clearly less than 1,000 mites. Often, one treatment in the autumn is not sufficient. Multiple treatments are necessary and these can be combined into one concept.

Monitoring the success of the treatment

The summer and autumn fall diagnoses serve to recognise a threat to the honey bee colonies in good time. In addition, they provide insight into the mite fall after each treatment in order to assess its effectiveness and to obtain information about the development of the contamination.

Are there more or fewer mites than last year? Are there differences between the colonies? Are these due to the beekeeping measures, the colony development or the treatments in the spring?

These and other questions about varroa infestation can only be asked if it is observed what happens in the colonies and kept track of the data. Dr. Liebig has been doing this every year for four decades. Without this intensive observation, the development of effective, tolerable and user-friendly treatment methods is not possible.

Counting mites takes effort. If you are not prepared to make this effort, you should at least estimate the number of mites that have fallen. This is less precise than counting, but it is certainly better than not having any understanding of the mite numbers at all.

When estimating, one must take into account that the mites have not fallen evenly distributed everywhere on the inspection board. The mites fall from the comb seams occupied by the honey bees and where the bees are closer together; in the centre of the honey bee cluster and from the comb seams where brood has hatched since the inspection board was inserted, more mites fall than at the edge. Sufficient accuracy is achieved if one marks a central comb alley and thus approximately 10% of the inspection board surface, counts the mites lying there and then calculates the total number taking into account the distribution of mites further on the inspection board.

Counting mites pays off. In the long term, it pays off if one can base the discussion on varroosis on the factors that influence their development, on their optimal control and on the associated problems. Only then is such a discussion also meaningful. Progress is only made if data are collected continuously.

Beekeepers' measures against the varroa mite

At the moment, only technical and chemical control methods are available. Biological processes - such as the use of natural enemies and diseases of the varroa mite or the use of attractants - have been and are still being sought, but so far without practical results. No valid progress can be seen in the effort to breed varroa-tolerant or varroa-resistant honey bees either.

Comparing the development of the contamination with the development of the population shows that the critical period for the populations falls in late summer when the size of the brood decreases drastically from the end of July and as a result the number of honey bees also shrinks sharply while at the same time the varroa contamination increases further. The relative degree of contamination of honey bees and brood then suddenly increases. The winter bee brood, which is reared from the end of August, is particularly threatened. This brood must be protected from an excessively strong varroa contamination. Only then can healthy long-lived winter bees develop, only then is a certain wintering of the honey bee colony guaranteed if it is also strong enough and provided with sufficient food.

Relationship between the number of fallen varroa mites found per day on the inspection board (in 3-7 days) and the number of mites in a starter colony or and older colony at different times of the year.				
period:	per fallen mite on the inspection board		treat when in	
	in overage colony	in starter colony	in overage colony	in starter colony
July & August	**x 150**	**x 150**	> 10 mites/day	> 5 mites / day
September & October	**x 300**	**x 300**	> 5 mites / day	> 1 mite / day
November & December	**x 500**	**x 500**	> 1 mite / day	> 1 mite / day

For timely and adequate protection against varroa, the use of medication is necessary. But this must not lead to residues in bee products!

This principle excludes the use of fat-soluble active substances; because these can settle in the wax when administered repeatedly and eventually pose a risk to the purity of the honey. This premise limits varroa treatment to the use of formic acid, lactic acid and oxalic acid.

These organic acids are water-soluble and therefore cannot contaminate the wax. They are also natural components of honey. Of course, they may not be used before or during the nectar flow or only after the last honey harvest.

In particular, the application of formic acid, which is introduced into the colony by evaporation, leads to a strong acidification of the honey that is left in the colony or the winter food provided for the treatment, which, however, has no negative effect on the development of the treated colonies and their wintering. In the acidified winter food, the formic acid content decreases to normal values during the wintering and in the spring. In addition, the winter food is eaten and the remaining winter food is removed when setting up the honey super.

With regard to effectiveness, organic acids are not miracle cures. Their effective application requires optimal conditions. For example, satisfactory treatment success with oxalic acid or lactic acid can only be achieved in brood-free colonies and with formic acid, which is also effective in sealed brood, only if it is warm enough.

Even under optimal conditions, the degree of effectiveness is rarely high enough that a single treatment is sufficient.

A modern varroa control that does not leave any residues in the honey bee products even when applied regularly, that is also applied in such a way that the colonies do not suffer from it, but remain healthy and capable, whereby the practitioner does not put himself in danger, presupposes a total concept in which the biology of the bee colony and of the varroa mite are taken into account and the measures to combat the parasite are coordinated. Then a fourth aspect can also be honoured: *as little as possible, as much as necessary*. Proven components of such a total concept are:

- the use of drone brood as a varroa trap
- the removal of sealed/capped worker brood for the production of brood layers and their subsequent treatment in a brood-free state with oxalic acid or lactic acid
- the summer- or late summer treatment of the production colonies after the end of the flow with oxalic acid (in the "SaT" concept) or with formic acid (in the "Complete formic acid treatment concept") and
- the treatments in late autumn and early winter ("residual mite removal") of the then brood-poor or brood-free colonies with oxalic acid

By removing sealed drone brood and worker brood and using them as varroa traps, the development of varroa infestation should be slowed down.

The treatments after the last honey harvest in late summer and autumn are intended to protect the winter bee brood from excessive varroa infestation.

Removing the mites in late autumn or early winter ensures that the overwintering colonies can start the following spring with as few mites as possible at the start of brooding, so that they do not build up a dangerously high varroa infestation until late summer.

The same goal is also served by treating brood-free splits in May/June. The young colonies that result from this should only be treated again after feeding for the winter. Treating young colonies in August with formic acid can lead to enormous brood damage, because the young colonies reach their brood maximum around that time and they are usually (still) in a brood box.

Cutting drone brood

The use of drone and worker brood as varroa trap is considered a technical control method. The sealed brood is cut out of the frame together with the enclosed mites and destroyed or treated and returned to the colony. A worker brood frame (with youngest brood) and honey bees can also be used to form a split with brood and bees, after which the treatment follows when that frame has become brood-free.

Removing and cutting out drone or worker brood cannot be called a biological control method, and certainly not if the brood is destroyed.

Removing sealed drone brood as a "varroa trap" is therefore worthwhile because drone brood is attacked more by mites than worker brood and mites multiply much more strongly in drone brood.

The time of sealing the drone brood lasts two to three days longer than that of the worker brood. As a result, the mites in the drone brood generate more offspring. If many mites are given the opportunity to do so for a longer period of time, the number of mites increases rapidly. While with exclusive multiplication in worker brood one can assume a monthly doubling of the number of mites, with multiplication in drone brood a tripling or even quadrupling is possible every month; of course as long as the honey bee colony is raising drone brood, which is usually the case from April to July.

The stronger infestation of the drone brood cannot be traced back to an active preference of the varroa mites over worker brood. The drone brood does not secrete a substance that particularly strongly attracts the mites.

The reasons lie elsewhere. Two of them are known:

- The time frame in which the mite females (can) penetrate the brood that is ready for sealing is twice as long for drone brood as for worker brood.
- The drone brood is clearly visited and fed more often by nurse bees in the larval stage than worker brood.

The apparent preference only becomes particularly visible when the drone brood is directly next to or at the same time as the worker brood and both are in the appropriate stage (shortly before sealing/capping). If a mite forced to lay eggs finds no drone brood, she will go into a worker brood cell without fail. One should also consider that in a normally strong colony there are many more worker brood cells than drone brood cells, on 2 brooded building frames there are usually 10 to 15 frames with worker brood cells.

If the trap is functioning properly, about 50% of the mites are hidden in the sealed brood. This means (with the reproduction of the remaining mites in the worker brood that with consistent cutting of drone brood the mite count remains approximately constant during the "treatment". But even a smaller proportion of captured drone mites would be reason enough to cut the drone brood before the drones and the mother- and daughter mites capable of reproduction with them emerge, because of the stronger reproduction.

The repeated cutting of drone brood also has another advantage, so that the use of the "building frame" is also worthwhile if one does not think of varroa mites. The building frame helps to satisfy the "building instinct" and the "brood care event" of the honey bees, which counteracts the emergence of "swarm mood". It is an excellent swarm mood barometer.

A higher effectiveness of the use is achieved if at least one "ready to catch" frame with open drone brood is permanently present in the colony.

The procedure of cutting the drone brood can of course only be applied from April to June/July. After that, the honey bee colony usually does not raise any more drones or only a few.

The recycling of the drone cell construction

The construction frame wax can be used as an additional wax production.

The construction frame can also be provided with frame wire several times lengthwise and crosswise to support the construction of the cells. The brood is then unsealed with a knife and knocked out or washed out together with the mites. The advantage is that the colony can then immediately be given a built-out drone frame that is immediately provided with eggs. Of course, one then renounces the wax production.

The washed-out drone brood can be used to feed fish or chickens (who will be than your friend forever). When washing the drone brood in a suitable double (honey) sieve, the degree of varroa infestation can also be seen.

Chemical treatment

A vast number of agents against the varroa mite are used in Central Europe. A few more are added every year, so a list would soon be incomplete.

The approval situation varies from country to country and also changes from year to year. Unfortunately, the approval of a product by the government does not immediately guarantee that it can be used without further ado and without hesitation, even if the instructions for use are strictly adhered to.

The use of chemical agents is always associated with risks that can of course be evaluated separately from agent to agent. This evaluation should not only take into account the treatment result, but also the effort, the tolerability by the honey bees, the formation of residues and the possible dangers for the practitioner.

Regarding these four points, the agents differ greatly in part. The residue problem was and is the most neglected. It is the distinguishing criterion whether the agent is suitable.

The formation of residues in the wax or honey depends on:

- the solubility of the agent (in fat or in water),
- the time and
- the quantity of the required number of treatments,
- the quantities used and
- the method of application.

Since the situation on the pesticide market is constantly changing, there is little point in presenting the various agents here. That is why it is limited here to describe the methods of application using a few examples.

The long-term contact treatment

The active substance is applied to a carrier that is hung in the occupied comb seams of a honey bee colony or is placed on the honey bee colony. The honey bees constantly absorb the active substance as they walk over the carrier. The honey bees constantly touch and clean each other in the colony. In doing so, the active substance is spread throughout the colony and on the combs.

When the carrier is present in the colony for weeks, the mites that emerge with the brood also come into contact with the active substance. If the treatment is carried out over a period of more than 12 days - that is the time of the sealing/capping of the brood - an effectiveness of 90% can certainly be achieved.

Strips or pieces of plastic, jute, wood, paper and soft fibres are used as carrier material. The plastic strips are industrially manufactured, the active substance is "processed into them". Thin strips of jute, poplar or balsa wood are soaked in an aqueous active substance solution and then air-dried.

To treat a colony, two to eight contact strips are required. The strips have a hole at the end for hanging and are hung with a nail in the separated comb seams of the brood space. They are removed after treatment.

It often happens that the honey bees seal or remodel the carrier material, so that already a few days after hanging, not enough active substance is released and the dose given to the honey bees is no longer effective.

The paper or soft fibre cloth soaked in the active substance is placed on the honey bee colony after drying and is then quickly demolished by the bees and removed from the hive. Depending on the weather conditions or the activity of the honey bees, the demolition (and thus the treatment) can take just a few days or a few weeks. In unfavourable weather conditions or after feeding, it may happen that the honey bees do not walk or process the material as intensively as would be necessary for a sufficiently high treatment result.

The long-term contact treatment only works optimally when the colonies are still fully active. The treatment should therefore be applied in late summer immediately after the end of the nectar flow and the removal of the honey super. Only then can damage to the winter bee brood due to a heavy varroa infestation be prevented.

The long-term contact treatment does not require much work. However, it has two major disadvantages:

1. With fat-soluble active substances and with repeated treatment - year after year - it leads to a huge load on the wax.
2. Mites that are resistant to the active substance soon appear. This is an experience that has been gained worldwide.

A long-term contact treatment is also possible with cloths or strips containing oxalic acid, but many tests have shown that this is not sufficiently effective.

Feeding

During the development of several products, it was tested whether a sufficiently high treatment success could be achieved by adding an active substance to the feed during feeding in August. The results were not very satisfactory. This applies to the administration of active substances in both liquid feed and feed dough. In addition, combating varroa with "beneficial food" easily leads to residues in the honey harvested later, regardless of the time at which this "beneficial food" was administered.

In 2018, the report that varroa mites can also be killed with various lithium salts that are used in human medicine as antidepressants if they are dissolved in sugar water and fed to the bees made headlines. This worked very well in laboratory tests with bees in cages. When treating honey bees at the beekeepers apiaries, the effectiveness on the varroa mites was relatively poor and the bee brood also suffered damage. Much more research is needed before a satisfactory result is achieved.

Evanescence

For control by evanescence, only substances are suitable that release active substances into the air at normal temperatures, which then act on the mites as contact or respiratory release. Essential oils, alcohol, ether and formic acid were tested. Of these substances, only formic acid and thymol have proven themselves as treatment agents.

The repeated application of thymol preparations leads to a smelly load on the wax, which in turn is a danger to the purity of the honey. There are a large number of dispensers for the application of formic acid.

Nebulisation

A nebuliser with a power source is used to nebulise emulsions with active substances. The emulsion with active substance is blown into the honey bee colony as an aerosol through the entrance or from above using a special attachment. The treatment of a colony takes one to two minutes.

More than the amount used for dripping is used for the active substance. The active substance introduced as an aerosol acts as a contact poison. Decisive disadvantages of aerosol treatment are that

- not only the honey bees
- but also the combs
- with the stored, sealed or open (!) food and
- the entire interior is treated and also
- the practitioner.

During the treatment of an apiary, the practitioner is exposed to a surrounding active substance cloud that settles on clothing and exposed parts of the body and is also inhaled.

Evaporation

Evaporation is still practiced with oxalic acid. There are two methods. In the older method, the oxalic acid vapour is "blown" into the colony through the entrance with an evaporator, of which there are many versions worldwide, which is associated with an abundance of "oxalic acid vapour" in the environment of the honey bee colony. The more colonies standing next to each other are treated one after the other, the longer the practitioner is exposed to the substance. Therefore, as with fogging, ample precautions must be taken.

In the second, younger method, the oxalic acid in the colony is evaporated in an empty container placed above it with an ingeniously constructed stove and a tea light. (That 'stove' can also be a spoon bent twice). The practitioner is not near the honey bee colony or at the apiary during the treatment and therefore, after the treatment, "only" has to deal with the residues of the oxalic acid that have settled on the walls of the empty container. In addition, the fire hazard must be taken seriously.

By evaporating oxalic acid, only the mites that are on the bees are destroyed. Therefore, a sufficiently high effect can only be achieved if the colony is completely brood-free. This is rarely the case in late autumn or early winter. However, this handicap is therefore relative because the treatment is tolerated very well by the honey bees and can therefore be repeated without harming the honey bees. With a block treatment in late autumn/early winter, honey bee colonies can be made virtually mite-free so that the next varroa treatment is not necessary until late autumn of the following year.

Spraying

A spray gun is required for spraying. A simple plant sprayer that can be operated by hand and with which the number of spray bursts can be dosed based on the honey bee population is sufficient. When using pressure sprayers that are used, for example, for weed control in the garden, it is easy to overdose.

Spraying is quite labour-intensive. The colony must be opened and each frame must be lifted. In addition, the practitioner must protect himself so that the spray mist does not come into contact with the skin or is inhaled. This also applies to the application of oxalic acid and lactic acid.

When using an oxalic acid solution, a combined treatment is possible in which the honey bee cluster hanging in the ground is sprayed from below and comb alley by comb alley is dripped from above. Spraying is very suitable for brood-free and closely spaced colonies in the summer.

Drip (drip from above)

Various products are available for the drip method. Most of them contain oxalic acid as an active substance and are also distinguished by different admixtures that, such as intensive frequent use has shown, are of no importance for an acaricidal effect. For a long time, it was believed that the active substances acted as a feeding poison when dripped. Adding sugar would then stimulate the honey bees to better absorption. But this assumption that the dripped agents work "systemically" was already rejected as incorrect in the first experiments that were connected to the research into the effectiveness of dripped oxalic acid. All drip agents work as contact poisons.

They stick to the moistened honey bees and are distributed in the honey bee cluster either by them through movement and body contact with other bees or by dust effects. These substances then come into effect as contact poisons on the mites that sit on the honey bees.

Therefore, all drip agents can also be used as contact strips or cloths, whereby the honey bees contaminate themselves with the active substance by walking on them or by gnawing and demolishing them.

With a treatment by dripping, a treatment result of 90% can be achieved, but only if the honey bee colonies are treated in a brood-free state, are also close together and the active substance reaches all the honey bees of the dripped colony infected with mites in the correct dose.

The even distribution of the active substance is less well guaranteed in the drip process than in the spray process. Furthermore, the honey bees that are directly hit by the drip can easily receive an overdose and honey bees that are further away are more likely to receive too little oxalic acid. As a result, it is better to spray the solution of the active substance into the bee seams with a thin jet than to let it fall on the honey bees in thick drops. Overdosing leads to bee mortality, underdosing to too low an effectiveness.

In addition, the active substances are only active for a short time on the infected honey bees. They are distributed less well in the honey bee cluster or are inactivated more quickly when the colonies are brooding, processing food or when there is a nectar flow. Therefore, they should only be used to treat brood-free and closely packed honey bee clusters.

The active ingredients

A treatment against varroa without residue problems is only possible with the organic acids formic acid, lactic acid and oxalic acid. Lactic acid is only used in a spraying process. In this case, it works less well than oxalic acid. Since oxalic acid has also been approved for treatment in the spraying process, only formic acid and oxalic acid are requested. Formic acid also works in sealed/capped brood and is therefore suitable for treating brooding colonies in late summer and early autumn, oxalic acid works in treating brood-free splits and swarms in spring and in removing the "residual mites" from colonies in late autumn or early winter.

The deficient effect in the sealed/capped brood is compensated for by measures in the colony care management which, as in the case of "Splitting and Treating", make the treatment of brood-free colonies possible and at the same time guarantee comb hygiene.

In the case of treatment by sublimation of oxalic acid, the advantage is that such a treatment is tolerated much better by the honey bees than dripping or spraying an aqueous oxalic acid solution. Sublimation also requires less effort and is user-friendly, if it is practiced from above in an empty container with a tea light and a small stove. With a block treatment in late autumn/early winter, the varroa infestation can be reduced to such an extent that the next treatment is not necessary until the autumn of the following year.

Formic acid

Formic acid has a worse image in beekeeping than oxalic acid, because many mistakes can be made when using it. Two extremes can be distinguished in the negative experience reports that have arisen in this way.

One says: formic acid does not work.

The other says: formic acid kills the queen, the brood and also entire colonies.

Both statements can be explained easily. If the dosage is overly cautious to completely exclude the loss of the queen, the brood or bees, then the formic acid often has no or insufficient effect because the hive air concentration does not even reach the limit of mite mortality. And vice versa, whoever applies formic acid according to the motto "a lot helps a lot" not only kills the mites but also his colonies. This can even happen if one believes that one has used formic acid according to the instructions.

No other agent for the treatment of varroa has so many application forms described to date as for formic acid. Neither the permitted nor the unpermitted forms of application do justice to all the requirements that arise, for example, from the various hive systems or from the expectations of many beekeepers ("it has to be simple"). The toxicity (and not only) of formic acid for mites, bees and brood depends primarily on the concentration of formic acid in the hive air. This hive air concentration can be influenced by many factors, of which the beekeeper has only a few control. These are the amount of formic acid introduced, its concentration or dilution and the way in which it is introduced. Other less predictable factors are the strength of the colony, the size of its brood, the availability of open food, the distance between the applicator and the brood nest, the type of hive, the hive size and the external conditions.

After application, the formic acid evaporates and mixes with the hive air, which the mites tolerate less well than the honey bees. However, with formic acid, the distance between mite mortality and honey bee mortality is relatively small. The threshold for honey bee mortality can be reached quickly. Exceeding it can have catastrophic consequences for the honey bee colony.

For the success of the treatment, it is not only important to achieve a sufficiently high hive air concentration, but also how long the formic acid remains in that concentration in the hive air. A treatment for a longer period of time usually has a better effect than a treatment for a short period of time.

Precautions when using formic acid

Formic acid is corrosive. It must not come into contact with the skin or be inhaled. Therefore, protective clothing (acid-resistant gloves and safety goggles) must be worn during all treatments and water must be kept ready.

Also be careful with contact with things that have been in touch with formic acid. Wear (latex) gloves here too.

Short-term treatment

With short-term treatment, the applied formic acid usually evaporates within a day.

The treatments with the sponge cloth and the "Illertic mite plate" (IMP) are "shock treatments". The applied formic acid evaporates suddenly, which is also reflected in the reaction of the colony. An audible humming sound rises. For example, if the dosage is incorrect at high temperatures, panic can break out, which in certain circumstances leads to the immediate exodus of the colony or even the loss of the queen. This can be prevented by applying cooled formic acid.

In the short-term treatment with the medicine bottle, this precaution is not necessary. The formic acid that drips from the medicine bottle evaporates very slowly at first, it takes a few hours before the hive air is filled with formic acid. The bees have time to get used to the formic acid and do not panic. Therefore, 85% formic acid can also be used at high temperatures so that a good effect is also guaranteed in the sealed/capped brood and the treatment does not have to be repeated.

Short-term treatments with the medicine bottle are therefore more suitable than treatments with sponge cloth or IMP.

Long-term treatments

With long-term treatments with formic acid applicators, uniformly higher treatment success is considerably easier to guarantee than with short-term treatments. However, not all applicators are equally good in terms of operation. This depends primarily on how the evaporation takes place. In this respect, there are differences among the applicators depending on their construction. With regard to the evaporation of formic acid, the large number of applicators can be divided into three groups:

- Type RI (rapidly increasing, then gradually extinguishing):
  Evaporation reaches its maximum very quickly after insertion or opening of the applicator ("shock treatment") and then gradually extinguishes as the applicator becomes empty. This group includes the Krämer plate, the Apidea acid applicator, the Liebefelder formic acid dispenser, the universal evaporator and the MAQS strips (Mite-Away-Quick-Strips). They work according to the same principle. The carrier material is soaked with formic acid and hermetically packed or covered. For the treatment, a small defined surface of the carrier material is exposed from which the formic acid then evaporates into the colony. The evaporation depends on the amount of formic acid that is released. As the formic acid is released the

applicator becomes empty and as it becomes empty the evaporation rate also decreases. This does not guarantee even evaporation. In addition, the formic acid in the carrier material attracts water during treatment, which also causes the concentration to decrease continuously. This also naturally causes the evaporation rate to decrease.

- Type SI (slowly increasing, then gradually dying out):
  The evaporation starts very slowly, reaches its maximum after a few hours and then gradually dies out, just like with type RI, as the applicator becomes emptier. Representatives of this type include the bottle or glass honey pot with wick and the Burmeister evaporator. The formic acid is poured into a glass or plastic container and evaporates via a wick that protrudes into the formic acid through an opening or a slot in the lid of the container. The formic acid does not evaporate as suddenly as with the ST type applicators because it takes some time for the wick to fill with formic acid at the beginning of the treatment. Only then is the maximum evaporation achieved, which nevertheless only lasts for a short time. Because with evaporation the container becomes emptier. The less deeply the wick protrudes into the formic acid, the smaller the amount that evaporates per unit of time.
- Type SI-RC (slowly increasing, then remaining constant):
  The evaporation starts very slowly, reaches its maximum after a few hours, as with type RI, but then remains constant at this level until the applicator is empty. The degree of evaporation is not dependent on the amount of formic acid in the applicator, thanks to the design. Type SI-RC includes four applicators: the Nassenheider evaporator vertical and horizontal, as well as the plate evaporator with the medicine bottle and drip outlet. The further development of the latter has led to the Liebig dispenser.

Honey bee tolerance

When using applicators type SI and SI-RC, the formic acid is tolerated more easily by the honey bees than with type RI, because the formic acid evaporates slowly in these types, giving the honey bees time to get used to the hive air and adjust to it. While a shock effect can occur when using type RI, this is avoided in any case with types SI and SI-RC. Applying or inserting an applicator of type SI does not necessarily lead to a panic reaction from the colony; in extreme cases, it can lead to an exodus of the colony or even to the loss of the queen. The colony usually growls and the honey bees quickly move downwards or upwards. When using an SI or SI-RC applicator, the reaction that can usually be observed is that the honey bees stay away from the applicator as long as the sponge is soaked with formic acid.

While the formic acid evaporates in the hive, the colonies ventilate more strongly at the entrance, regardless of which applicator is used. Plants growing directly in front of the entrance are affected by this.

Brood(in)tolerance

A long-term treatment clearly has an effect on brooding behaviour, regardless of the type of applicator used. Two effects must be taken into account:

1. The queen stops laying eggs or she does lay eggs but the maggots that emerge cannot tolerate the formic acid in the hive air and die. Brood that is already present can also die and be cleared away by the honey bees.
2. The sealed/capped brood continues to develop normally during the treatment. They are only at risk when they hatch. The young bees that emerge die if they are exposed to formic acid without protection. This is the case if the emerging brood does not have enough distance from the applicator or - in weaker colonies - is not protected by the bees. Both effects can be reduced if the long-term treatment is only carried out after feeding for the winter and not before mid-September. At that time, the colonies themselves already largely limit their brooding activity. A long-term treatment with formic acid ensures that the colonies leave the brood earlier and faster, which is not detrimental to the wintering strength and does not have a negative effect on the wintering of the colony.

Treatment success

With applicators of the SI-RC type it is easier and sufficiently higher to achieve a treatment result than with those of the SI and RI types because the effective formic acid concentration is maintained in the hive air as long as the applicator still contains formic acid. Of course, this is only the case if the sponge (the blotting paper) is set large enough, taking into account the population size, the hive type and the hive volume. With the applicators of the SI and RI types, the formic acid concentration in the cabinet air depends on the filling level. That is why it decreases over time. As a result, the effect also decreases.

Handling

There is no great difference in handling between the applicators. With the newer applicators of the RI type, the carrier material is provided with formic acid, then the evaporation surface is set and the applicator is placed on or in the colony. Dr. Liebig reports that reuse is made more difficult because the applicator placed on top is cemented by the honey bees. In particular with the Apidea acid applicator, the three parts slide into each other much more smoothly than when first used. Nevertheless, these applicators show a clear improvement on their predecessor, the Krämer plate.

User safety

The requirements for user safety are met more by the dish evaporator with the medicine bottle than by any other applicator, also because the application of highly concentrated formic acid to the applicator does not occur at the honey bee colony. There is no need to fiddle with formic acid. Remove the screw cap and place the medicine bottle with the drip outlet turned upside down on the prepared blotting paper. That's all. The treatment can be stopped at any time without danger and without much effort. Remove the bottle, turn it upside down and close it again with the screw cap. The remaining formic acid can remain in the bottle and be stored for the next treatment. With the other applicators this is not possible or only possible under certain conditions.

The Nassenheider evaporator vertical, reports Dr. Liebig, is used mounted in an empty frame and this frame is hung next to the honey bee seat. The subsequent models are placed in an empty container on the honey bee colony. In the vertical and horizontal models the storage space is separated from the evaporation space. The storage space, closed on all sides, is connected to the evaporation space, open at the top, above an opening located directly at the bottom. Through the opening in the partition wall, only enough formic acid leaks into the evaporation space until the liquid level in the opening closes.

This ensures that the sponge (the blotting paper) is always immersed to the same depth during the entire treatment period, regardless of the amount of filling in the storage space, and therefore also evaporates almost the same amount of formic acid.

In a more recent professional model, the formic acid slowly, continuously and completely runs out into a bath filled with sponge cloth and evaporates there completely in the empty hive box. In the event of a sudden change in the weather, where it first becomes cool and later very warm, it can happen that too much formic acid evaporates from the filled bath in a short time, the bee colony leaves and the brood dies.

The medicine bottle

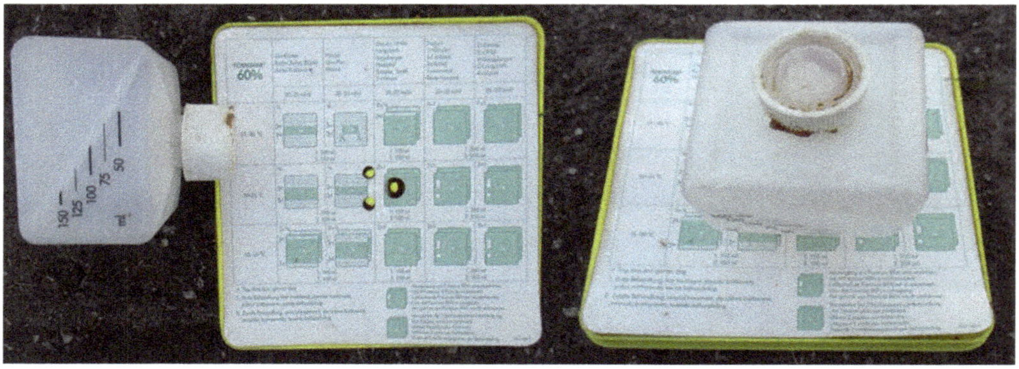

The Liebig dispenser. When the bottle is empty and the tissue paper has dried, the paper is nibbled on and off.

The treatment with the medicine bottle was already extensively presented in the description of the care in late summer. An empty box is needed for the treatment. The medicine bottle with formic acid is placed upside down on a dish in which a thin sponge [a blotting paper] is placed. The liquid level in the dish does not exceed the edge of the bottle opening. Only as much formic acid flows out of the bottle as evaporates via the blotting paper in the colony. The evaporation performance is set via the blotting paper surface. It is marked on the blotting paper how much surface of it would be needed for a treatment of your colony.

LD (Liebig Dispenser) three days treatment			
	Perrenial colonies	(on two boxes)	Young colonies (on one box mostly)
Formic acid 60%	140 ml		70 ml
Formic acid 80%	100 ml		50 ml

Oxalic acid

Oxalic acid is a metabolic product that can be detected in the blood and urine of humans and animals either as a free acid or as a salt. Oxalic acid is also widespread in the plant world. Particularly high levels of oxalic acid are contained in wood sorrel, sorrel, spinach and rhubarb (and also parsley, berries, beans, beets, nuts and cocoa also contain a lot of oxalic acid). Oxalic acid is also a natural component of honey.

Oxalic acid is used in industry and trade, among other things, to remove rust and ink stains and is processed in metal polishes, to mordant wool, to print acid dyes on wool, to bleach leather and straw braiding, in photography for the production of developers and colours and as an additive in cosmetic products (hair care products).

For handling oxalic acid, the professional associations prescribe various precautionary measures that are generally customary for handling etching substances.

They range from hygiene measures (do not store food in the work environment, nor eat, drink, sniff or smoke; store work clothes and leisure clothes separately; change and clean contaminated clothing) via technical and organisational safety measures (only work in ventilated areas, the work area must be equipped with a washing facility, provide an eye wash or eye wash bottle) to personal protective measures (eye protection with safety glasses, hand protection by wearing acid-resistant gloves, skin protection [apply protective ointment to all exposed areas of the skin]).

In beekeeping, oxalic acid is used to combat varroa. For about 40 years, the spraying process was the standard method for (post-)treatment of brood-free colonies in late autumn or early winter.

The laborious treatment (each frame has to be lifted and sprayed) and the associated dangers for the operator are probably the reason why this method has found few followers among beekeepers.

A "career leap" for oxalic acid was triggered by the introduction of the drip process, which was developed in Italy twenty years ago. It was particularly impressive that the mites fell down after treatment with the "Italian mix" of water, sugar and oxalic acid in the ratio 10:10:1, which corresponds to an oxalic acid concentration of 8%, without any dead honey bees falling down on the mesh bottom/floor or in front of the entrance. This led users to believe that the drip process would also be well tolerated by the colonies. Very many beekeepers worked according to the method "a lot helps a lot" and dripped their colonies several times and thus to death; because repeated dripping with oxalic acid leads to the colonies flying themselves hive-empty. The loss of colonies or the weakening of the colonies after the oxalic acid treatment by dripping caused the euphoria in the beginning among the beekeeping population about finally having a simple treatment method that leaves no residues in the honey bee products, quickly changed into a negative attitude towards oxalic acid. The drip method only gained acceptance again when it became clear that in brood-free colonies an adequate effect with good honey bee tolerance can already be achieved at an oxalic acid concentration of 3.5%.

In the early 1990s, the 'Fischermühle' (Fishermen's mill) proposed a third treatment method by evaporating dry oxalic acid in the gauze bottom, which, like the spraying process, had its origins in Asia and which brought about renewed progress in beekeeping, although it had no advantages over the spraying process. The complexity of the device and the time required are greater. The protection of the operator requires adequate safety clothing from head to toe. The initial assumption that the evaporated oxalic acid would also work in the sealed brood turned out to be incorrect.

A treatment rate of 90% can only be achieved in brood-free colonies under optimal conditions. For each of the three processes - spraying, dripping and evaporation - the "optimal" conditions must be defined differently. This also applies to a fourth method, the oxalic acid treatment with strips or cloths. These are hung in the honeycomb seams of the honey bee colony or placed on the occupied frames. When walking over or nibbling off the strips or cloths, the honey bees take up the oxalic acid crystals and distribute them in the beehive. The longer the strips or cloths remain in the colony, the longer the effect lasts. Of course, it has not yet been possible to achieve a certain and sufficiently high treatment success with this method at all times in late summer or autumn. The optimal conditions or circumstances are not even remotely known yet.

Evaporation of oxalic acid

For the treatment of honey bee colonies with oxalic acid by evaporation, there are a multitude of high-tech devices that blow the produced vapour into the honey bee colony from above or through the entrance and thus from below. Their applications are impressively demonstrated in numerous videos that can be found on YouTube. Of course, there are no data on the effect, time frame and number of necessary treatments and on the honey bee tolerance anywhere.

The method described below has been tried and tested for years by Dr Gerhard Liebig and Dr. Pia Aumeier and there are still some questions open and not yet satisfactorily investigated.

This method has so far only been tested in wooden hives in Zander size and with Zander boxes as mounted empty boxes. In plastic hives it could possibly get too hot and the plastic could start to melt due to the burning tea light or the stove. Then the lid gets a big hole. Under certain circumstances it could lead to fire. There is also a risk of fire in wooden hives if flat boxes are used as empty containers and the lid does not have enough distance to the stove. It might help if in both cases the metal roof only serves as a lid.

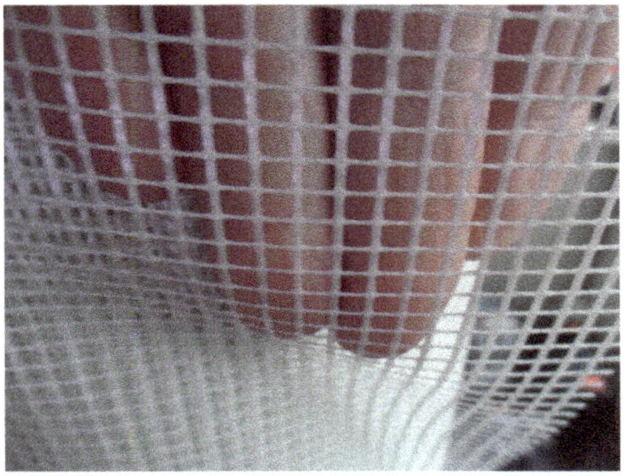

The most convenient is the so-called spoon stove: a spoon bent twice, fixed by a tile and a tea light underneath and on a plasterer's mesh, with approx. 2 grams of oxalic acid powder/time/colony on the spoon is the recipe (See also [again] p. 195.

To secure the combs for liquid and hot and possibly also burning wax that can escape when the tea light bursts, the various constructions offered on the market can be placed on a fireproof base (a saucer or something similar, possibly sprinkled with moist sand). There are also small stoves made of stainless steel with a base and four closed walls that are approximately the height of a tea light and are extended "airily" towards the top. A small container is placed on the stove in which the oxalic acid is placed - in powder form or as a tablet - and is placed above the burning tea light for treatment.

The treatment does not work in sealed brood. Only the mites that are on the honey bees are tackled and not all of them. The degree of effectiveness on the "bee mites" can vary from <50% to >90%. It depends on the weather conditions and the hive climate. A satisfactory effect on the "bee mites" is achieved in cool or cool-humid weather (= no flying bees during the day) and the treated colony has no or only a few brood. These conditions are not encountered before late autumn or early winter.

In the summer, evaporation of oxalic acid works very poorly in the colonies, even when they are brood-free, such as in the resulting splits with the queen that arise during "Splitting and Treating" and the "brood colonies" that have become brood-free after 21 days after the division. In summer weather conditions, the degree of effectiveness is well below 50%. Evaporation of oxalic acid is also not recommended for the treatment of brood-free colonies in the winter when they are huddled together at freezing temperatures. It works clearly worse than a drip treatment.

Therefore, the treatment period is clearly limited to late autumn /beginning of winter before it gets really cold. Because of the brood still present at that time, a block treatment is carried out, in which the evaporation is repeated at intervals of 1-3 weeks until almost no more mites fall down.

The oxalic acid is evaporated from above with the help of a tea light in a stove in an empty box that is placed on the brood box(es). In addition to this empty box, which is usually already on the colony in late summer care as an evaporation space for the formic acid, the following is also required:

- plasterers' mesh (plaster gauze / reinforcement mesh): it completely covers the bee seat and prevents the honey bees from accessing the empty box; because they would be able to extinguish the tea light by forming clusters. The mesh size of the plasterers' mesh may not be more than 4 mm. Suitable plasterers' mesh is available in the hardware store (or via a web shop, e.g. roll of 50 cm x 15 m, so ideal for a co-operative buy). A 1 m wide 10 m roll provides material for 48 colonies, if these are in a "single hive". The mesh is cut to the outside dimensions of the boxes. This ensures that there are no holes above the bee seat through which the honey bees can end up in the empty box. The empty box is then placed on top of the densely applied plasterers' mesh.
- a tea light. Tea lights are of different quality. The most expensive are not necessarily the best. Tea lights with thin wicks go out more easily. You can prevent the going out (somewhat) if you let the tea light burn "above without" for a few minutes after lighting it. In windy or rainy weather, you place it in the empty box and cover it "airily" with the inner lid or the metal roof. There are tea lights that burn out without any problems.
- a stove with a saucer underneath: Up to now, various stoves have been tested by Dr. Gerhard Liebig and Dr. Pia Aumeier. The "airy" stoves with a thin roof made the best impression. When using compact stoves, it often (more often) occurs that the oxalic acid applied does not evaporate completely. This can - but does not always have to - lead to a poorer effect. Such events have led to the conclusion that the amount of 2 g oxalic acid hydrate used world-wide for evaporation through the entrance is not necessarily necessary to achieve a satisfactory effect. 1 g seems to be sufficient for colonies on one box. More research is needed for an explanation.
- crystalline oxalic acid hydrate, 1 tablespoon, teaspoon or measuring spoon for dosages of 2 g (for colonies on two boxes) or 1 g (for colonies on one box). The

oxalic acid can also be placed on the stove in tablet form.

- acid-resistant gloves,
- a lighter.

Method of application

The treatment is carried out in the evening after the honey bees have flown. See for the manual with pictures: page 195.

The entrance is not closed or remains reduced.

1. After the drop diagnosis, a clean inspection board (inspection board) is inserted. In order to achieve a high degree of effectiveness with "evaporation from above", it is not only not necessary to close the entrance, but also not necessary to insert the inspection board. In corresponding tests, the inspection board was only inserted after the oxalic acid had completely evaporated. After the treatment "without the inspection board", the same number of mites fell as with the treatment "with the inspection board", measured by the mite fall during the next treatment. This result also will be investigated further by Liebig/ Aumeier.
2. Put on acid-resistant gloves.
3. The colony is opened and covered with the plasterer's mesh. Care must be taken to ensure that no honey bees are trapped on the upper moldings of the frames.
4. The empty box is set up. No bees or wasps may be in it. This must also be taken into account just before putting on the inner lid.
5. The saucer or tile is placed on the plasterer's mesh.
6. After the holder with the oxalic acid is placed, the tea light or the tea light holder with the tea light is placed. The tealight is lit and the empty box is immediately sealed with the inner lid and the metal roof. The foil is of course not stretched under the inner lid when sealing. Otherwise it would start to melt during the treatment and fall down onto the hot stove.
7. The oxalic acid starts to evaporate shortly afterwards. The evaporation takes a few minutes.
8. During evaporation, oxalic acid vapour escapes from the leaky areas of the box, including the entrance. The vapour plumes dissolve after a short time. When there is wind, they are blown away. In the evening half-darkness, the plumes are hardly noticeable.
9. During evaporation, oxalic acid crystals settle on the plasterer's mesh and on

the inner walls of the empty container, most of them on the inner lid directly above the stove.

These residues are particularly easy to detect the next day and later. Residue often also collects on the saucer. Their formation can be explained as follows: the oxalic acid vapour rising during evaporation condenses when it hits the cool inner lid. Crystals hanging from the lid directly above the stove form. These grow and fall down when they have become too heavy and thus collect on the cool saucer. This residue does not come into effect.

It may also happen that part of the oxalic acid on the stove does not evaporate, even if the tea light burns completely. This unevaporated oxalic acid often remains on the stove as flaky crystal glitter. In both cases, the treatment success is not significantly affected. It must be suspected that oxalic acid also condenses on the inner walls of the empty container, on the occupied and unoccupied frames, in the mesh bottom and on the inserted inspection board and from there dissolves in the hive moisture that has settled there.

The empty boxes and tools are not removed until the aim of the block treatment, to make the colony as mite-free as possible, has been achieved. Depending on the varroa infestation and the brood size, the treatment must be repeated more often. The fewer mites that fall after a treatment, the later the treatment must be repeated. The result of the mite fall, which is determined with a mite fall after a treatment of 1-4 weeks, serves as a decision-making tool. The block treatment can be concluded if the mite fall caused by a treatment clearly decreases from treatment to treatment and less than 100 mites have fallen after the last treatment. In that case, further treatment can be omitted.

After the last treatment, the tools and empty boxes can be put away and cleaned with water (don't forget to wear gloves). It does not harm the colonies if they are overwintered with a set up empty box and above an open mesh bottom box. Then the cleaning of the tools and empty boxes only follows in the spring.

Evaporating oxalic acid from above with a tea light and a stove in a set up empty container. The honey bee colony is covered with a dense plasterer's mesh. The tools and the empty container are only removed after the last treatment.

Mite fall

The treatment causes a clearly increased mite fall on the first day. This increased mite fall continues for more than a week. Most of the mites that died after this or because of this short-term treatment fall in the first week, the maximum is on the second or third day after the treatment. After that, the mite fall decreases clearly. The mite fall that continues for days after a treatment is not caused by the oxalic acid remaining in the colony for days. As with spraying and dripping, the evaporated oxalic acid is also only active in the colony for a short time (<1-2 hours). The majority of the affected mites are not killed immediately, but only damaged and then fall off the honey bees later. Some can even step into the brood, die there and when this brood emerges approximately two weeks after the treatment, cause a long-term increase in the fall of dark mites. If after the treatment clear mite stages can be seen in the fall, then brood has emerged after the treatment and with it the other mites that survived the treatment in the sealed brood cells and were able to multiply further. If the colonies are still brooding in late autumn and are heavily infected, a quick repeat of the treatment is advisable. A further treatment should be planned if more than 100 mites have fallen on the first day after the treatment.

Treatment Protection

When using oxalic acid, do wear acid-resistant gloves. This safety measure applies to both the preparation and the after-treatment. The contaminated tools and empty boxes must be rinsed with water. Here too, (therefore) wear gloves.

Do not inhale any vapours during the treatment. It is even recommended not to be present at the apiary during evaporation with a stove (curved spoon) and a tea light.

These instructions are easy to follow. In principle, the tea light is lit after all steps to install the stove properly have been taken and the oxalic acid has been placed as well. Take care that never oxalic acid is sprinkled on an already heated stove c.q. after the tea light is lit. If the tea light has been burning for a few minutes, evaporation will follow quickly. After the start of the treatment, do not stay near the honey bee colony or at the apiary. A few metres' distance is sufficient to observe, if wanted so, the vapour development without danger.

When treating multiple colonies, first prepare all the colonies, then light the tea lights and close each colony immediately. Pay attention to the wind direction when determining the order in which the colonies will be treated. Work against the wind....

Thymol

In addition to the organic acids formic acid, oxalic acid and lactic acid, thymol is also considered an "alternative" pesticide against varroa. It was tested in the 1990s at the Hohenheim Institute for combating varroa during rearing.

In Europe, three thymol preparations are on the market and approved. The oldest is the product *Apilife VAR*, originating from Italy, which, in addition to thymol, also contains the essential oils camphor, eucalyptus oil and menthol, of which of course only thymol has a satisfactory effect against varroa. Apilife VAR consists of plates of vermiculite that are soaked in the above-mentioned oils and must be placed on top of the honey bee colony for the treatment and must remain there.

They are worn down by the honey bees.

Another product is *Apiguard*, in which thymol is included in a special gel composition. The thymol is placed on a dish on the honey bee colony and evaporates in a few weeks, more at the beginning than at the end of this period so that at the beginning of the treatment the mites that are on the honey bees are exposed to a stronger dose of thymol than the later mites that emerge with the brood. This also applies to the other thymol preparations. It is therefore advisable to repeat the treatment after about three weeks.

Thymovar is a third thymol product. A sponge cloth is used as carrier material, two of which are placed on the honey bee colony and covered with the foil, the inner lid and the metal roof. After two to three weeks the colony receives a second dose.

The success of thymol treatment depends very much on its evaporation and therefore on the temperature. In the cold season thymol remains virtually ineffective.

In addition, the distance between the place of placement and the place of the bee seat plays a role in the degree of effectiveness. In colonies on one box it is easier to achieve a satisfactory treatment than in colonies that sit on two boxes. The treatment with thymol can also have a damaging effect on the brood. In the immediate vicinity of the place of placement (approximately a hand's breadth) brood (and previously stored food) are completely cleared out. if it concerns sealed/capped brood, pupa parts appear in the fall on the inspection board, which are noticeable when inspecting the success of the treatment. The pupal legs are particularly noticeable. In addition, the laying capacity of the queen is clearly reduced. When laying eggs, she keeps a considerable distance from the thymol preparation that has been applied. Due to the effect on the brood, it is advisable to start the treatment only after the first feeding and to apply the thymol preparation in such a way that there is a large feeding wreath between it and the brood nest. Clearing out the feeding cells directly under the laying place has no negative effect on the further development of the honey bee colony and its hibernation.

The colonies can be fed during the treatment with thymol.

Due to the duration of the treatment, the mite fall continues for a very long time, even if most mites fall in the week after the start of the first or second treatment. It takes more than a month for the natural mite fall to be established again. This makes it difficult to assess whether a follow-up treatment in late autumn is still necessary. This applies in particular to colonies that are wintered in two boxes. The treatment result can vary widely in these cases, rarely reaching 90%, even if they are treated with a double dose as for colonies in one box.

For colonies in one box, the effect of a two-part treatment is usually above 95%.

Thymol is fat-soluble. During the treatment it is stored in the combs, which serve as a "source of smell" until spring. The combs of colonies treated with thymol still smell intensely of thymol after the spring inspection in March. The smell is only lost when temperatures rise, which is probably also contributed to by the expansion of the warm brood nest. A treatment with thymol in spring or throughout the year with a "thymol frame" that is hung in the colony during the season, unmistakably leads to residues of thymol in the honey. It is therefore not recommended. If honey tastes of thymol it cannot be marketed. Thymol has no place in honey either. This applies in principle to all agents, including preparations that contain formic acid, oxalic acid or lactic acid.

Apistan

The active ingredient in Apistan is Tau-fluvalinate. Originally it was licensed by a firm in Ireland, but in the mean time also a Chinese producer/seller occurred. However, as with previously used acaricides, the development of resistance to fluvalinate could be expected, especially since it attaches to the wax and is therefore continuously present in the honey bee colony in low concentrations. Also a Dutch study, published in 2002 (Tjeerd Blacquière, Christ Smeekens, Bram Cornelissen, Rob Smits, Jeroen Donders, Jan van den Eijnde, Jeannette van der Aa en Sjef van der Steen, PPO Bijen, Hilvarenbeek, *Resistentie van varroamijten tegen Apistan al op veel plaatsen in Nederland*, maandblad voor imkers -monthly magazine for beekeepers-, juli/augustus 2002, p. 218-219) has confirmed that resistance of varroa against Apistan has indeed occurred and spreads. So better not use it (anymore).

Comb hygiene, comb stock and wax processing

With young colonies, comb hygiene is not a problem. They contain one or two (at most three) older combs in the bottom box and nine or eight frames with young construction that were given in May as foundations or as empty frames with wire and were built up in the time that followed with continuous feeding.

When the young colonies have become very strong, they can be given a second box for rearing, which is also filled with young or at most once or twice incubated combs. A few foundations can also be included. These are then quickly built up with rapid rearing.

With older colonies that are on three boxes in August, a rejuvenation of the comb construction should also be sought. To this end, when reducing the size, the bottom box, which is usually brood-free in August, is removed; the dark combs are sorted out and melted. This should be done with all multiply incubated combs and only clear unincubated combs should be stored in the frame stock box. These do not need to be protected from attacks by the wax moth. In brooded frames, however, protection against wax moths is absolutely necessary.

In this respect, one should refrain from using chemicals. In particular, one should be warned against the use of paradichlorobenzol-containing agents to combat wax moths. They are indeed easy to use, they are placed in the closed and frame-filled "box towers" where the active substance evaporates continuously and kills the wax moth larvae, but as a fat-soluble substance they leave residues in the wax.

And that is less pleasant, because these residues easily pass into the honey when the combs (or the wax that has been extracted from them) are back in the colony and are used to store the honey.

With regard to residues, the use of acetic acid or sulphur discs is less problematic. Of course, such treatments must be carried out twice or three times (at intervals of 3 to 4 weeks), because only the larval stages of the wax moth(s) and not the eggs are killed.

However, it can also be done completely without chemicals.

The combs are stored cool, airy and dry so that they do not become mouldy. The combs are sorted beforehand. This is worthwhile because the brooded combs are more likely to be infested by wax moths than unbrooded ones. They lack the vital protein that wax moths ingest with the webs, honey bee droppings and pollen residues. The simplest measure against moth damage is therefore to store only clear unbrooded combs. Then you do not have to worry about wax moths. Brooded combs are usually infested by wax moths that enter the storage area as eggs with the hive or comb material, hatch there and go in search of food. A chemical-free defensive measure is to regularly check the stored combs, approximately once a month, especially during the warm seasons or when stored warm. The complexity increases with the number of combs.

Webs are easy to recognise and can be removed with tweezers or a dissecting needle. The infected combs can also be placed in the freezer for two or three hours. This time is sufficient to kill larvae and any remaining eggs of the wax moths at a temperature of approximately minus 20°C. The cold treatment can also be carried out prophylactically. Of course, the boxes in which the brooded combs are stored and the cover must also be placed in the freezer for two days, because wax moth eggs could also stick to the inside of the boxes.

An alternative to the frost treatment in the freezer could be the heat treatment in a warming cabinet. A refrigerator converted with a suitable heater and thermostat is suitable as a warming cabinet. You could use such a cabinet also to liquefy crystallised honey. A box with ten brooded frames also fits in it. The heat treatment against wax moth eggs and larvae is successful if it is carried out on such a cabinet for at least three hours at a temperature of 48°C.

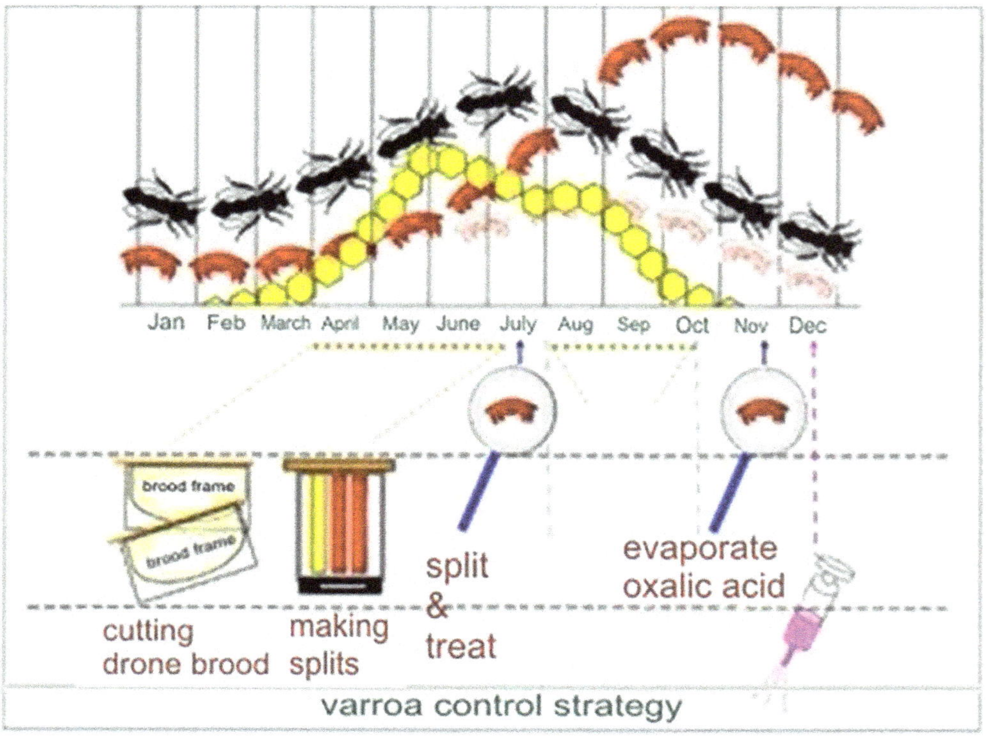

after an image of Dr. Otto Boecking, LAVES Institut für Bienenkunde, Celle (Germany)

Caring for honey bee colonies as concise monthly considerations

January

If you have prepared your colonies for the winter in the right way, nothing needs to be done to the colonies. Check occasionally,

- whether the colonies are still standing as they were during the last visit
- whether the metal roofs weighed down with stones are still on the hives after a storm
- whether a entrance has become blocked. If it is wide open and the mouse mesh blocks the access for the shrews, field mice and house mice, it cannot become blocked.
- whether a woodpecker has been at work, which does this especially after snow has fallen.

When checking during the year, it is advisable to have a camera at hand.

February

Also in February there is little or nothing to do with the honey bee colonies. After a mild winter one must be careful that the food supply of the colonies does not run out prematurely. During the "food check" just lift the colony from behind. Than open the colony that is the lightest in weight if it is found to be too light. The "too light" is confirmed if no or little sealed/capped food can be seen from above and next to the bee seat when "looking from above" (under the foil placed on the box). Then the frames are lifted and the food supply, if present, is accurately estimated by means of the "eighths method" (Liebefelder estimation method).

If there is too little food, a frame exchange can help. If there are no clean feeding frames of more than heavy colonies, because no colony has too much food, one can definitely also start using feed dough, because that can be provided without the use of a feeding device. A thin slice is simply placed on the colony under the foil and covered with a turned-over lid that then shows its concave side downwards. Small portions of less than 1 kg are sufficient and the feeding is repeated if the flow is still delayed. Water is needed to process the feed dough. Therefore, there must be flying weather during the day.

Another feeding variant that can be done without lifting frames and where the colonies do not have to fetch water is liquid feeding from below. For this purpose, a flat container or (plastic) dish filled with sugar water or syrup is placed on the mesh bottom directly under the bee cluster. The free height of the mesh bottom box determines the depth of the dish. Depending on how long and wide it is, a bowl can hold up to 3 liters of liquid.

A handful of dry leaves, small cut branches or coarse wood chips can serve as a swimming and climbing aid.

Even at very cold temperatures, at just a few degrees above zero, a cluster of honey bees hanging from the comb seams will be on the swimming and climbing aid within no time after the food has been put down and will empty the bowl with impressive "team work". In normally strong colonies, the bowl is empty and dry after a night and the cluster of honey bees has retreated back into the comb seams.

Caution is also advised when feeding from below. Robbery can break out in flying weather. A previously reduced entrance helps against this.

March

A practice is to overwinter the honey bee colonies - old colonies as well as young colonies - in two brood boxes. The old colonies are used as production colonies in the spring and almost all of them should have been provided with a new queen after the care in late summer and autumn. The young queens required for this are obtained by uniting young colonies or also via the "Split and Treat (SaT)" procedure. With the colonies overwintered in two boxes, the first intervention is not until April with the placement of the honey box.

That is why you would rarely have anything to do with the colonies in March. You only have to make sure that the colonies have sufficient food supplies and keep an eye on the awakening of the plant world that is oriented towards the climate and the weather.

Cold waves in the spring only slow down the development of the colonies if they last longer than a week. Then there is a slight lack of pollen. the larvae are no longer fed, they are "recycled", eggs and young larvae first. This leads to a "brooding break" after which a "bee break" follows two-three weeks later.

If according to the weather forecast there is flying weather during the day and there is no more freezing cold at night, the mouse wire is removed so that the colonies can work the fallen dead bees out of the hive and the fallen ones do not accumulate in front of the mouse mesh.

The population strength is estimated on the basis of the number of occupied comb seams. The assessment data are recorded in an apiary protocol (hive card) and later processed at home in for example an Excel table in which the notes of the further data of the population checks from April to December are also included.

Only in colonies that show little air traffic is a closer inspection necessary, in which the frames are lifted and the brood nest is assessed. Very weak but otherwise intact colonies are placed smaller on one brood box and with a few feeding frames placed on a queen excluder at a stronger colony. The colony placed above receives care support from below. With the help of the strangers, the queen of the upper box quickly creates a brood nest that is not inferior to the brood nest of the queen in the lower box. These double colonies are separated again in April.

April

Shortly before the start of the first mass flow that begins with the cherry blossom, the honey box is placed on top of a queen excluder. But first, a small frame in the brood chamber is lifted out to create space for an empty frame, a construction or building frame. This frame is placed in position two or nine. If two or three frames still contain a lot of winter food, these are also removed and replaced by two or three foundations. These foundations are each hung between two brood combs. There they are quickly built out and covered with eggs in a large area. In this way, the brood nest in the upper brood chamber is considerably widened and the honey box as such is more easily accepted.

The honey box is provided with six or four clear frames as a "comb core" and 2x2 or 2x3 foundations on both sides. Those who do not have clear frames only use foundations.

The procedure "setting up the honey box" starts with tilting the upper brood box for each colony. The smoker is only used after the colony's reaction to the opening of the hive has been assessed. The following ratings are given for this purpose:

1 = "No honey bee moves, the colony behaves very calmly".

2 = "The bees walk in the comb seams, the colony remains calm".

3 = "The bees storm out of the comb seams, the colony is restless".

4 = "The bees fly up, the colony is very restless".

5 = "The bees fly up, a few attack, the colony is poisonous".

If a sting is given, a 5 is given without further ado. If a lot is stung, the colony gets a 6. These behavioural evaluation figures should be recorded on the hive card and later processed in for instance an Excel table. Based on the data collected there, a potential "breeding queen" is selected later in the late summer.

In principle, it is actually advised to work in the colonies without further protection, but always keep the smoker within reach and do not shy away from using it. The same counts for the veil.

Approximately three weeks after the hanging of a construction/building frame, the sealed/capped drone brood created in it is cut out. During this intervention and during the following swarm checks, the behaviour is assessed and "scored" twice, the first time after the removal of the honey super and the second time during the subsequent "tilt check", during which the play cups created in the upper box, which become visible from below when tilted, are also carefully inspected whether they will become queen cells (soon). The smoker, reading glasses and an LED flashlight ensure a free and accurate view into these play cups.

After successful population development, this intervention, the second in the year and still in April, is also used for the bloodletting of strong colonies. In the case of a more hesitant population development, bloodletting is only done in May. Each colony supplies a brood frame with honey bees sitting on it, but without the queen!. Nine of these brood frames with honey bees on it and a feeding frame form the basis for a foster colony of "class". On at least one of these brood frames, there must also be the youngest brood present for the production of queen cells. The foster colony is placed outside the flight radius of the brood frame providers.

In April, the remediated weak colonies that were placed above a queen excluder on a strong colony four-five weeks beforehand are also moved. On the new position, they are given a honey box on the brood box and a little later the second brood box from below is added.

With this, the double colony can also separate without having to move. Then the lower strong colony is put aside and the upper one is placed where the strong colony was just now. All the flying honey bees of the double colony now fall to the 'upper colony'. That is why it must immediately be made wider from "one to three". The moved colony also gets a building frame and a box that will be the honey super.

May

Caring for honey bee colonies during the swarming period is part of the High School of Beekeeping. With measures to prevent swarming and, if these do not help, to prevent swarming, the colonies must be kept from swarming in such a way that the production capacity is not reduced. There are many recipes for preventing swarming. After thorough research of Dr. Liebig into intermediate floor splits, brood distancing, double colony beekeeping in various variations, salto beekeeping and various bloodletting methods (suction box, heating box, floating box, split box, split wth a queen, splits with brood and bees it has become clear that to bloodlet once early (towards the end of April/beginning of May) and "softly" (with one or two brood frames with honey bees sitting on them) and during the swarming period (from the end of April/beginning of May to the end of June/beginning of July) to consistently conduct a swarm check every seven days. This measure prevents swarming and slows down the spread of varroa infestation as well. The removal of drone brood is integrated into this seven-day rhythm. Those who have not used drone brood as a varroa trap in late summer will have at least twice as many mites in their colonies and will have also a lower harvest.

During the tilt inspection, play cups with larvae are easy to recognise. With reading glasses on your nose and an LED flashlight in your hand, even play cups with already eggs will not go undiscovered. The uncertain beginner can practise by carrying out a tilt inspection every four days.

If swarm cells are discovered - one is too many! - all brood frames of that hive must be taken in hand and examined carefully after shaking off the honey bees. It does take time (about five minutes per colony), but it must be done.

Lifting frames and breaking cells is only necessary for colonies that have created swarm cells. Even if these colonies have been in a swarming mood for a long time, breaking swarm cells is less difficult than other methods of preventing swarming. And: it has no negative effect on colony development or collection production.

From years of experience, it appears that on average one third of the colonies in an apiary get into a swarming mood.

Preventing colonies that want to swarm in time can seen as a competition for you: each colony is checked every seven days from the end of April to the beginning of June. If you give up you lose.

If you lose, there are consequences: the swarmed-off colony will not produce honey for weeks. The swarm that has flown off is doomed if it is not caught. When catching swarms, there is a chance of an accident. These three things should be enough as motivation to "persevere" the swarm control every seven days.

In May, the population increase is also carried out in four steps according to a time schedule. On day x, a bloodletting practice is carried out and a collective brood layer is formed. On day x+9, the queen cells are broken, the breeding frame is larvaed and hung in the middle of the foster colony. When breaking cells, the foster colony is placed more spaciously with a box added from below. If there are no built-out frames with some food available for this, the box to be placed is provided with foundations and a closed feeding frame with syrup, whereby the provision of swimming and mounting aids must not be forgotten.

On day x+19, the hung and accepted queen cells are caged with three-five guide bees. If the cells have expanded too much and do not fit in the cages, they are carefully trimmed with a warm pocket knife.

On day x+21, all queens and all worker brood from the foster colony have emerged. This is used to form fertilisation colonies. Each fertilisation colony consists of a "honey bee frame" (with >1000 bees), a young queen and an extra feeding frame. The half-filled edge frames from the honey chambers of the production colonies are very suitable for this. If more queens have emerged in the foster colony than sufficient strong fertilisation colonies can be formed from the bee mass from the foster colony, the missing brood-free "honey bee frames" are taken from the stronger production colonies. In these fertilisation colonies, the young queens are placed in a box with feeding dough.

The fertilisation colonies are placed four at a time in a box with a four-part bottom (quarter bottom). Three partitions divide the box into four. Each part has its own entrance (on each side of the box).

When disbanding the foster colony, the brood-free "honey bee frames" are sprayed with oxalic acid before they are used for a fertilisation colony. This guarantees a varroa-poor start. The next varroa treatment will then not take place until September or later.

As a breeding material supplier (from which the larvae for the queen cells are obtained) a colony is chosen that has shown peaceful behaviour in the previous year. That colony will therefore not have received a new queen in October of that year.

June

In June, when regularly carrying out swarm checks, it is important not to neglect to harvest the spring honey and to care for the young colonies at suitable times. The spring honey harvest takes place when the rapeseed bloom is over. This recommendation also applies if there is no rapeseed "in front of the door".

The honey is only harvested if it contains less than 18% moisture. This objective is achieved if one

(1.) only reaches for the sweeper (the hand broom) or the bee exit after the end of the harvest and
(2.) immediately after a rainy day and
(3.) brings in the honeycomb early in the morning in dry weather.

The ripeness of the honey that has not yet been sealed/capped is tested with the splash test. As soon as the test proves negative (the falling of a drop of honey from the comb is enough) the honey harvest must be suspended, even if more than two-thirds of the honey in the honey box is sealed/capped.

The honey is always moistest on the wall frames. If a second honey box is set up during the blossom period, the honey in the upper box is moister than the one below. Honey matures with age, but of course only in the colony. As soon as the honey is harvested, there is a risk of loss of quality.

The easiest method to obtain sufficiently dry honey is: "wait and see". Of course, you should not wait too long. Glucose-rich honey from dandelion or rapeseed tends to crystallise quickly. Such honeys must of course be harvested and extracted as soon as possible after the flow, before they become stuck in the combs.

How quickly the honey loses water after the visible end of the flow is impossible to predict. Many factors play a role in this. A few examples:

The heavier the comb, the moister the honey. In "thick" combs, honey is moister and dries more slowly than in "thin" combs.

The colonies that are positioned in the shade produce on average drier honey than the colonies that are positioned in the sun.

Very good blossom flows result in a higher water content. If the supply or collection of watery nectar is very high for days, the collected nectar in the hive is also processed very quickly and sealed/capped too moist for the beekeeper. The maximum shelf life for the honey bee colony is assured because the honey can still release water in a sealed/capped state. Although of course less quickly than open honey. In principle, one should always assume that the water content of sealed/capped honey is more likely to be above 17% than below. Therefore, the honey should only be harvested when all doubt as to its ripeness can be eliminated; even if it is possible to dry the honey in a short time in a moisture-removing climate room before extracting it.

After placing a bee escape early in the morning, the honey box can be removed almost bee-free the next evening. The bee escape only works when beekeeping is done with a queen excluder and there is no brood in the honey box.

During the honey harvest, honeycombs are often damaged. This, as well as replacing the removed full honeycombs with empty frames, can lead to search flights by (other) bees and provoke robbery. The still weak young colonies are particularly threatened. That is why the entrances must be kept small. Work in the colonies that could lead to searching for food must in principle always be done in the evening.

Caring for young colonies

The young queens of the fertilisation colonies that are housed in a four-quarter bottom box are ready for the mating flight one week after their emergence. This takes place in fine summer weather. The mating flight can therefore be determined with the help of the weather forecast. Two weeks later, without major disturbance of the young colony, it can be reliably assessed whether "everything is in order". Then, when looking from above in the comb seams, sealed/capped brood can be seen. Colonies without sealed/capped brood are usually queenless and are united with the neighbours: pull out the partition and pull the frames together.

After another week the young colonies start to grow, first slowly and then faster and faster. Before it becomes too cramped in the quarter sections, they are moved to a brood box and moved to their "final" position. In the new home the young colony is hung against a side wall and immediately gets two extra foundations. In the new location the entrance is (also) kept small. A food supply to size guarantees undisturbed development. One or two containers filled with sugar water or syrup are placed as far away from the entrance as possible. Don't forget to add "vegetables" as a swimming and exit aid. A new foundation is added with each refill.

July

In July, after the lime blossom, the summer honey harvest takes place, where the same procedure is followed as for the spring honey harvest in June.

After the summer solstice the days get shorter. The honey bee colonies, old and young, all react by shrinking the brood nest, the stronger older colonies a little earlier than the weaker young colonies. From the end of July to mid-September the queen of an older colony reduces her average daily egg production from more than 1500 eggs to less than 500. How does the queen notice in the pitch dark of the hive that the days are getting shorter? How does she know that winter will start "soon"? It is probably a decision that is made and followed through by the colony as a whole. It is worth remembering that none of the workers in the colony that have to make the preparations have experienced "winter" and that none of them will experience "winter" either!

Winter bees only come into the world in the autumn. Their origins are similar to those of swarm bees. Newly emerged bees are hungry. They eat their fill of honey and bee bread, digest the food, produce feed juice that they cannot get rid of, because just like in May, a high care potential meets a low care requirement. The feed juice is deposited in the abdomen as a fat protein storage. And the winter bee is ready.

If a honey bee does not feed larvae, it lives longer, but of course only if it was born healthy and was spared from varroa infestation in the pupal stage.

Therefore, the varroa infestation in the colonies must not exceed the damage threshold during the breeding of the winter bees (= in the autumn).

The varroa infestation can be assessed on the basis of the natural fall of varroa mites via the "mite fall diagnosis". The precipitation of three-seven days on an inspection board inserted into the mesh bottom box and with the aid of a head magnifier. With each count, the number of mites found is noted in the apiary protocol (hive cards) and later entered for example in an Excel file where the mite fall per day is then calculated.

The treatments against varroa are integrated into the care in late summer and autumn. Two proven concepts are available for the production colonies. It is best to start with the concept "Split and Treat" (SaT) at the end of July after the harvest of the summer honey. Two mesh bottom boxes (floors) and a second cover (foil, inner lid and a metal roof) are required for each colony. To repeat again:

On day x, the colony is divided into a split with the queen and a queenless "brood colony" (split with brood and bees). There should be flying wheather on day x+1.

During the splitting, the colony is first separated. The honey box is removed and put aside. The upper brood box is "parked" on an inverted metal roof, the lower brood box is placed on a new mesh bottom box, the entrance of which has already been made smaller. Now the honey box is placed on the old mesh bottom box, which remains in place. This is where the split with the queen is created.

Then the queen is searched for. She is usually in the upper brood box on a brood frame.

She comes with a few helper bees in a queen cage that is closed with food dough. This cage is placed on an occupied comb alley in the (former) honey box. Then the brood colony is put together again and placed on top of or a few meters besides the split with the queen for the flying of. The already reduced entrance points in the same direction as that of the split with the queen.

About two-thirds of the honey bees from the shared colony end up in the split with the queen, where it takes one-two days for the queen who has been eaten free to start laying eggs on the clear combs of the original honey super. On day x+2, the split with the queen can be treated with oxalic acid against the varroa mite. The entrance is then also made smaller, because by the time the brood starts to emerge in the split with the queen, about half of the honey bees in it will have died.

Dripping works (in the summer) clearly worse than spraying, which is more difficult because then each frame has to be lifted and sprayed on both sides with oxalic acid solution.

That is why you could combine spraying and dripping by "spraying from below (sfb) - spraying from above (sfa)". The honey bee cluster hanging in the mesh bottom box is gently sprayed with oxalic acid from below with a sprayer after tilting the box. Then the same solution is sprayed with a syringe from above into the occupied comb seams. For this "sfb-sfa" treatment, approximately 100 ml of oxalic acid solution is used when the split with the queen with a honey bee cluster in the mesh bottom box is sagging and occupies nine comb seams.

The "brood colony" is placed next to its brush on day x+2 if not already done and if there is room for it. If not, the "brood colony" remains on top of the split with the queen until reunion and is first taken off each time it needs to be treated or fed and then put back on it later. That is often the case with most beekeepers.

The "Split and Treat" also includes the diagnosis of the fall on the inspection board (inspection board) before and after treatment. A book should be kept about the mite fall.

August

August is usually a period without nectar. Therefore, the activities on the programme must be carried out in such a way that no robbery breaks out: Only feed in the evening. Keep the entrances of weak colonies small.

At the end of the month on day x+21, the treatment of the brood-free "brood colony" is on the agenda for the "SaT" colonies. First, the "brood colony" is narrowed down to one box, the older combs (from the bottom box) are removed for melting down into wax.

In the evening or the next morning, the "brood colony" is treated just like the split with the queen 19 days earlier. The "brood colony" then clearly loses more mites than its split with the queen before, because it had retained most of the mites when splitting the colony and many mites were still able to multiply in the brood.

Feeding is only started when it is certain that the queen from the queen brood cell is laying the "right eggs". She does this with a probability of >90%.

The second concept for preparing the colonies for wintering is the "late summer care in four steps": reduce - treat - feed - treat. This rtft- or four steps concept is almost 20 years older than "split and treat". The production colonies were kept from April to August on two boxes with a queen excluder and a honey super. In the brood space, the rapidly shrinking brood nest since July has been moved upwards. After mid-August, there is no or very little brood in the lower brood box (with the older combs). That is why you best start the "late summer care in four steps" in the third week of August.

When reducing the size, the honey box is first removed and placed to one side. Then you tilt the upper box and inquire about the position and size of the brood nest. If there is still brood in the lower brood box, it is located there under the middle of the brood nest above.

The upper brood box is removed and placed to one side.

Then the lower brood box is removed and placed slightly diagonally on the inverted metal roof. When converting this box, the size and position of the honey bee cluster hanging out below it are observed. If it is very small or even missing, the colony is reduced to one box and initially cared for as a colony on one box.

If it indicates that there is still brood in the bottom brood box, a frame that "gives away" this is lifted. The widened comb alley that is created in this way allows a view of the neighbouring frames. If there are still brood frames in the bottom brood box, they are hung on the edge of the top brood box. For this, the edge frames - which are not yet brooded and which usually contain some food (honey and/or bee bread) - are then exchanged for the former honey box. The clear empty frames that remain after this "frame exchange" are placed in the frame storage.

Then the colony is put together again:

The top brood box is placed on the mesh bottom box and the honey box on top of that. Then an empty box is placed in which the honey bees from the bottom box are shaken off. These two-year-old combs are collected bee-free in a box and later melted down.

It should be emphasised that the hanging of brood frames is an exception. It usually does not occur if the reduction takes place in late August.

The empty box that has been set up is then used for the formic acid treatment. This treatment is carried out or initiated "in jerks" as usually:

- insert a clean inspection board,
- set the dispenser with a piece of blotting paper placed on it,
- screw the medicine bottle that has been filled with 100 ml formic acid 80% (140 ml formic acid 60%) and that has not been cooled with the drip outlet on it,
- make sure that the drip outlet is in the neck of the bottle (and not in the cap),
- press the bottle over the dispenser,
- close the colony with the inner lid and the metal roof.

The foil is folded up on top of the inner lid or placed under the metal roof during the treatment.

The treatment is only started if the weather forecast indicates that it will be warm in the summer (T_{max} > 20°C) during the treatment. Then the bottles that have been set up will empty in three-four days. If this emptying is expected due to the weather conditions, one starts feeding the feeders with feed syrup after four days. The inspection board is removed from each colony and the mite fall is counted. Approximately 20% in mites of the calculated mite count from the mite fall diagnosis of the colony should have fallen. Then the treatment tools are put away (wear acid-resistant gloves) and the foil is placed back on the colony, but in such a way that a gap remains open through which the honey bees can enter the empty box, a plastic storage box is placed on the foil and, depending on the food supply in the combs, filled with up to 12 litres of feed syrup. On top or in the syrup respectively comes a swimming and climbing aid "nature" (plant stems, twigs, grass, etc.: the "vegetables")

When the inspection boards are slid back in, so that the total mite fall can be observed by the formic acid treatment, is the next hive visit scheduled after three-five days. Then the food intake is checked and refilled where more than 12 liters of syrup has to be converted into winter food.

September

In September you may take care of three "population types" (A, B and C).

A. The production colonies that are cared for after the "late summer care in four steps" and that are on two boxes. They receive the last feed ration(s) and are then treated with formic acid for the second time.
B. The sub-colonies that have emerged after "split and treat" and that are on one box and perhaps still on top of each other, presumably next to each other. The " split with the queen " and the "brood colony" of a splitted colony together brood more winter bees (and also more varroa mites) than a non-splitted production colony. The sub-colonies are fed in September so that they have stored sufficient (= not too little and not too much) food after their reunion in October.
C. The young colonies that have grown steadily since June with good care have built out six-eight foundations and fill a box in September. That is sufficient. After the last feeding in September, they are treated for the first time with formic acid against the varroa mite if necessary.

Colonies on one box should enter the winter with 12 kg of winter food or six full feeding frames. The food is given in two-four portions over a period of two weeks around mid-September. As feeding vessels, dishes that can hold at least five litres are suitable and are placed in an empty box that is placed on the colony. If it is on the programme to unite colonies in one box in October, then a plastic feeding vessel with the width of two frames is hung on the edge of the hive is preferred by most beekeepers. Until the uniting with a second colony on one box, both are then on eight frames.

With colonies on two boxes, one does not have to worry that their space requirement for a brood nest and the winter seat will be limited by massive feeding. They can get the necessary food in one portion and earlier (after the first formic acid treatment). For this purpose, a plastic stacking box is placed in an empty box that is placed on top. The colony is covered with a foil but so that a gap is left open at the back (or front) through which the bees can obtain the food given. Also the partly colony covering foil prevents wild building in the empty box.

If the food is not taken in, the distance between the honey bee seat and the feeding vessel in the empty box placed above is too great. In such cases you can make an aid by "laying a track" and hang a cloth soaked with (a very small amount of) feeding solution over the edge of the feeding vessel without feeding solution dripping into the mesh bottom box. Or a plastic frame feeding vessel is hung next to the honey bee seat. Two frames must then disappear at that spot. These are hung in the empty box for the time being during feeding.

Honey bees cannot swim. Therefore, swimming aids (a thick layer of grass, corks, pine cones, twigs: the "veggies") must be provided on the food. In the case of feeding barrels with vertical and smooth walls on the inside, climbing aids are also required (a cloth hung over the edge or a bunch of twigs placed in the barrel).

Feeding gifts provoke robbery. Therefore, food is only provided in the evening. The entrances, and in particular those of weak colonies, must be kept small.

October

During the summer, the number of colonies has grown. It is reduced by uniting.

In the case of the sub-colonies that are cared for according to the "SaT" concept, this is done by simply placing them on top of each other. The (younger) queen of the set-up "brood colony" usually survives.

If this is to be guaranteed, the other queen must be caught. Best is to catch both from the colonies, mark the queen that must remain in the united colony (she gets a drop of colour on her back) and place her in a cage with a feeding dough closure on top. Before that, you can also change the boxes. In addition, the "brood colony" with the older combs is placed at the bottom and the "split with the queen" with the younger combs is placed on top. If the united colony does not have sufficient food supply - both sub-colonies still have brood, the size of which is determined at the same time when searching for the queen - a feeding frame remains in the upper box that is filled with syrup (and swimming and climbing aids). The search for queens for the "reunion" of a "breeding colony" (if it is placed above) with its "split with the queen" (which is placed below) proceeds as follows:

In the "breeding colony" that is still above, a frame is lifted at the edge, then the central frame that is made accessible by moving the adjacent frames en bloc. Then the central frame is lifted and searched for the queen. If she is not there, the next frames are next. The young queen is picked up by the thorax or the wings with the fingers of the right hand (if you are right-handed), changes hands between thumb and index finger, is marked with a colored pen or other marker in the right hand and then imprisoned in the prepared cage. After adding about three helper bees, the cage is temporarily put aside. Keep it nicely warm.

Then the shifted and lifted frames are put back in their old place. Then the box of the "split with the queen" is removed from the bottom and after inspecting the seat and the strength of the honey bee cluster in it, which may have to be driven into the comb alley with smoke, the box is placed on an inverted metal roof. In its place comes the box of the "brood colony" equipped with ten frames and covered.

Then the old queen of the split with the queen is searched for. First a small edge frame is lifted and put aside, then the central frame after shifting the adjacent frames en bloc... The old queen is killed or imprisoned for freezing. Biology teachers in particular are grateful customers.

Finally the box of the split (without the old queen, if she has been removed) is placed on top of the box of the "brood colony" and the cage with the young queen and her helper bees is placed on top of it, above that the foil, the inverted inner lid and the metal roof. The reunion is complete.

Perhaps there will also be feeding once. For this purpose, a feeding frame is hung in the upper box on the edge. Two frames will then have to make way. They can be parked in an empty box placed on top, which is also needed in late autumn or early winter as an evaporation space for the oxalic acid evaporation treatment. See below under 'November'.

The change of queen in the colonies that are cared for according to the "four steps concept".

Only after the hive cards have been assessed will a selected brood queen(s) remain in the colony. All other queens from older colonies are replaced by younger queens. First, the queens of the young colonies are caught, marked and caged with a few helper bees and some feed dough. Each queenless colony is placed on a young colony with a queen immediately after the queen has been removed. The cages with the young queens are kept warm, for example in the breast pocket of a shirt or behind the waistband of the trousers.

Then follows the exchange of the older or production colonies: The old queen is caught and the cage with the new queen and her helper bees is hung or placed in or on top of an occupied comb alley above the brood nest to be eaten free.

If the older colonies are only on one box, they are strengthened by placing a young colony on top of it.

After the queen change, all colonies are on two boxes and enter the winter with a young queen. The only exception is the "breeding queen". If the weather forecast announces nights with frost, all entrances that have not been reduced in the meantime are provided with mouse meshes.

With a diagnosis of the fall, the varroa infestation is determined. Whoever observes knows and is protected from surprises. The results are noted on the hive card and recorded at home in the (Excel) file of the colony in question (see also the example on page 95.

In good weather, ivy and mustard are collected and pollen is introduced. Such days are used to note the flight traffic of the colonies at the entrance and to think about the differences between the colonies on an apiary. The introduction of pollen has no negative effect on the wintering.

November

In late autumn/early winter, the colonies are freed from their mites by means of a treatment by oxalic acid evaporation. The observation of the mite infestation is also connected to the treatment. How often and at what distance the treatment should be carried out depends on the varroa infestation and the brooding condition of the colony. If the colonies continue to brood for a long time and have many mites (the natural mite fall is > 5 mites/day), the treatment is carried out more often and at shorter distances (of approximately 7-14 days). If the colonies have fewer mites and brood little, the treatment is carried out only two or three times at a distance of two-four weeks.

For treatment, an empty box is placed on a bee-tight plasterer's mesh on the colonies. An adequate stove is placed in the empty box in which a tea light is lit that evaporates the oxalic acid powder or oxalic acid hydrate tablets that have been deposited on top of the stove (see again page 191).

In contrast to the many high-tech devices that are used worldwide and with which the oxalic acid is evaporated through the closed entrance, the practitioner of the thermos sublimation with tea light and stove does not stay at the apiary when it evaporates from the colonies. All colonies on an apiary are treated simultaneously. That goes also much faster than one after the other.

The deposited residues in the empty box are removed after the last treatment with a damp sponge cloth (or a tuft of -wet- grass) when the empty box and stove are cleared away. Do not forget to wear gloves...

December

In December, the oxalic acid evaporation block treatment is concluded. About two weeks after the last treatment, the inspection boards are taken out for the last time and the last mites are counted. The mite numbers are noted on the hive cards and from there included in the relevant (Excel) files for further analysis.

Succes with bee better beekeeping!

Guidelines for box, mesh bottom box, lid, wooden roof and three-frames-box construction

Brood box construction [honey box = brood box]

As a guideline for the construction of a box, we will take the same method for making a three-frames box, because in principle the construction process is the same. Only no wire mesh is attached to the bottom of the box. The top and bottom remain open. The bottom of a box stands (for a start) on a mesh bottom box of about eight centimeters high that does have a wire mesh bottom and room for an inspection board underneath, see the respective guideline.

For one box, one panel/plank of 200 x 25 x 1.8 cm is sufficient.

If you want to build a complete beehive, you need at least three boxes and then three planks as indicated.

You also need approx. 26- screws (3.5 x 30 mm) and some wood glue.

As for the tools you need: a saw (machine), an (electric) screwdriver and a tape measure.

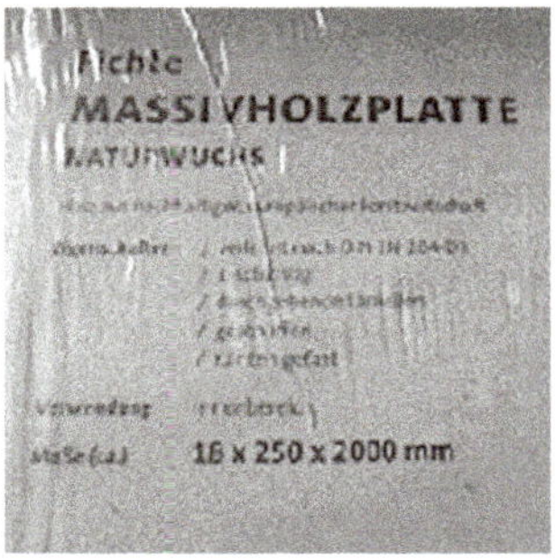

For one box, the plank is first sawn *** into
2 pieces of 52 cm length and
2 pieces of 38 cm length (you then have approx. 20 cm left)
If you build multiple boxes at the same time and therefore need multiple planks, you can also follow a different sawing procedure. It is a matter of calculating what is most economical in terms of residual wood. The starting point for each box is the number and size of the frames that are going to be used. Here, with the mentioned sizes, beekeeping with frames in (modified*) Zander format with so-called Hoffmann sides** has been assumed, 10 pieces per box.

* 'modified' means here that the top bar of the frame is not the regular Zander format (there are different beehive sizes, a Zander box or Liebig box is our choice as the most practical), but is made in the (thicker top slat) Langstroth format. As a result, the bees build less comb on top of the top bars of the frames.
We also keep bees with honey boxes that are identical to the brood boxes.
The boxes are interchangeable and the brood and honey frames as well.
Great advantages and convenience served.

** Hoffmann sides give the bees space in the right distance between the frames in the hive. In the Netherlands, for example, beekeeping is usually done in savings boxes in which frames with 'smooth' sides are used in the brood boxes. As a result, spacer strips - usually made of metal - must be placed in these boxes to create the right bee space. It is not possible then to move several frames all at the same time.

*** If you do not have a good or handy saw available, an alternative is to have the planks cut at a hardware store.

Length of the frames from below the top slat to the bottom: 21 cm

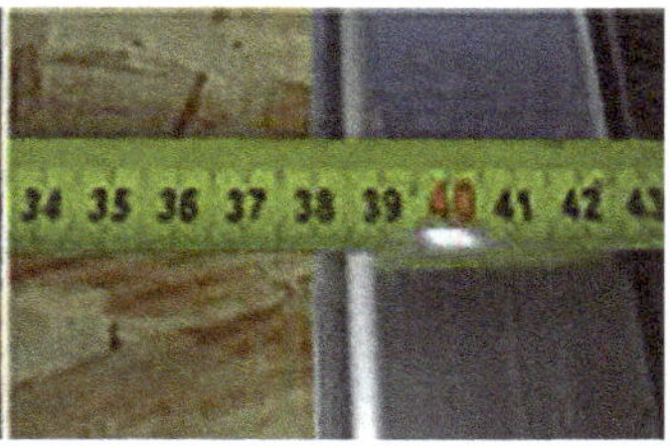

Pieces of 52 cm and 38 cm length (and another 25 cm width = plank width)

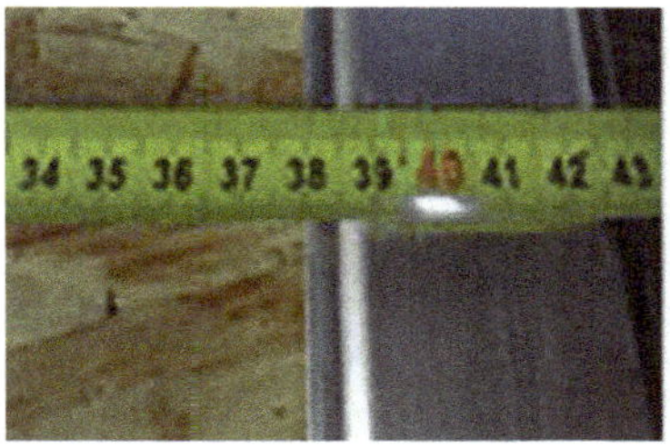

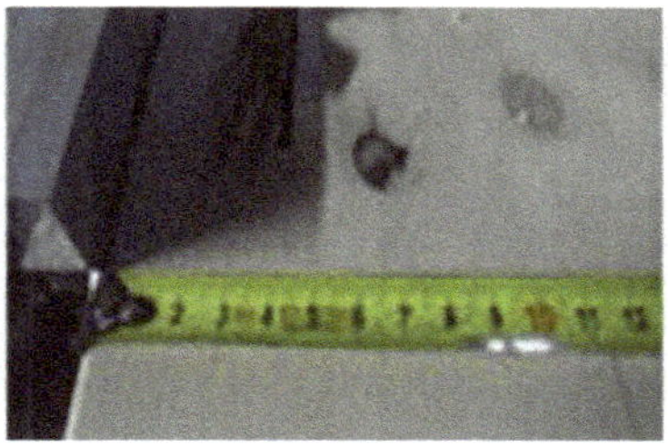

Then 4 cm is sawn off the width of the 38 cm wooden parts, so that these planks measure 38 x 21 cm.

The sawn off wood remains are also used later!

2 cm is sawn off from the width (25 cm) of the 52 cm wooden parts, so that we get planks of 52 x 23 cm.

These sawn off pieces are also used again!

Now all the planks (except for the two long remaining slats that still need to be shortened to 38 cm) that are needed have been sawn to the right size:
two x [52 x 23 x 1.8 cm] and
two x [38 x 21 x 1.8 cm] and
two x [38 x 4 x 1.8 cm]

You will also need approx. 26- screws (3.5 x 30 mm, flat head), some wood glue and a screwdriver

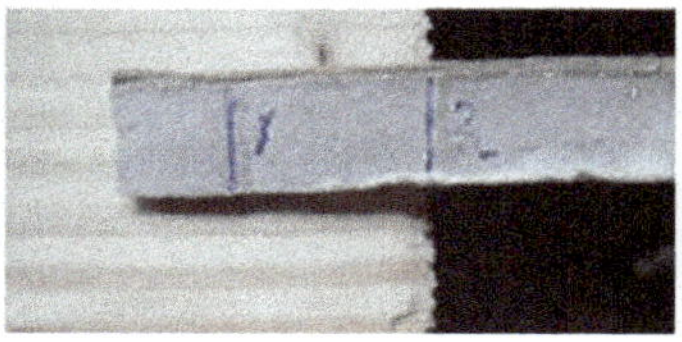

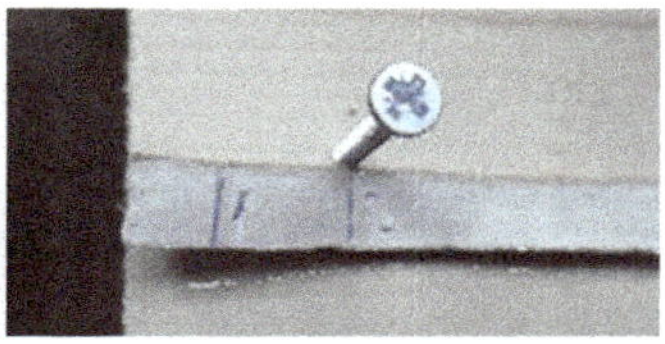

It is worth noting the dimensions on a strip of cardboard of the places where the screws need to be screwed in. So a line at 9 mm (1) and one (2) at 18 + 9 = 27 mm, if the thickness of the plank (panel) is 18 mm.

Now we start making the front and back wall that are completely identical. We take the two pieces of wood of 38 cm length and 21 cm height and also the two slats of 38 cm length and -at least- 4 cm width and the two narrower slats of 38 cm length.

A front or back wall consists of one larger and two smaller pieces of wood that are glued and screwed together.

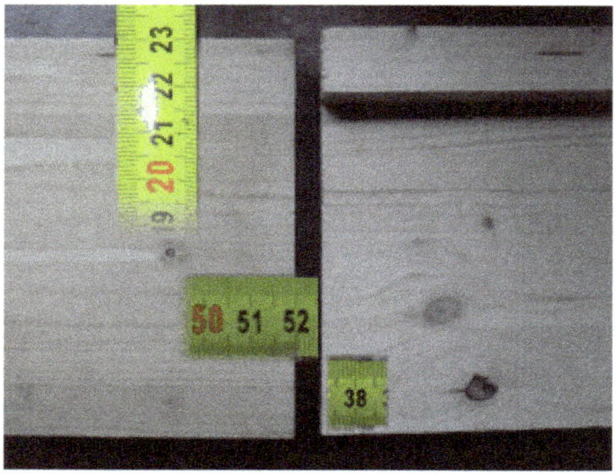

In principle, the assembly of the shelves is the same as when making the three-frames hive.

The assembly of the 38 x 21 cm planks and the slats of approx. 4 cm
››› from 21 to 23 cm high.

with wood glue and screws (3.5 x 30 mm, with flat head, best stainless)

You can use the remaining narrow slats, shortened to 38 cm, for the outer bottom of the side panels. Although slightly wider slats look perhaps a little nicer.

<< All the planks are then ready to be glued and screwed together

Make sure that the four planks are attached to each other in such a way that the box does not wobble when it is (on the ground or) on another box.

As a hive,two, three or sometimes four boxes are placed on top of each other and it is not wise if there are gaps between the boxes through which the bees can pass. Especially in late summer, this attracts robbery and the colony can easily be plundered.

just glue and screw and ....

the box is made !!!

(side of the box)

This is what a box looks like that is made entirely of one plank, as described above. This box can be a brood box, but also a honey box. It is much easier and also more efficient beekeeping if all boxes (and frames) have the same size.
If you want to use different slats at the top and bottom of the sides of the box for comfort or other reasons, or if you find this necessary, then of course you should look for the slats of your choice.

Mesh bottom box / Lid / Roof

The manual according to Dr. G. Liebig:

The wooden parts of the mesh bottom box and their dimensions for a LiebigZander box.

In any case, the dimensions must correspond to the (brood and honey) boxes used.

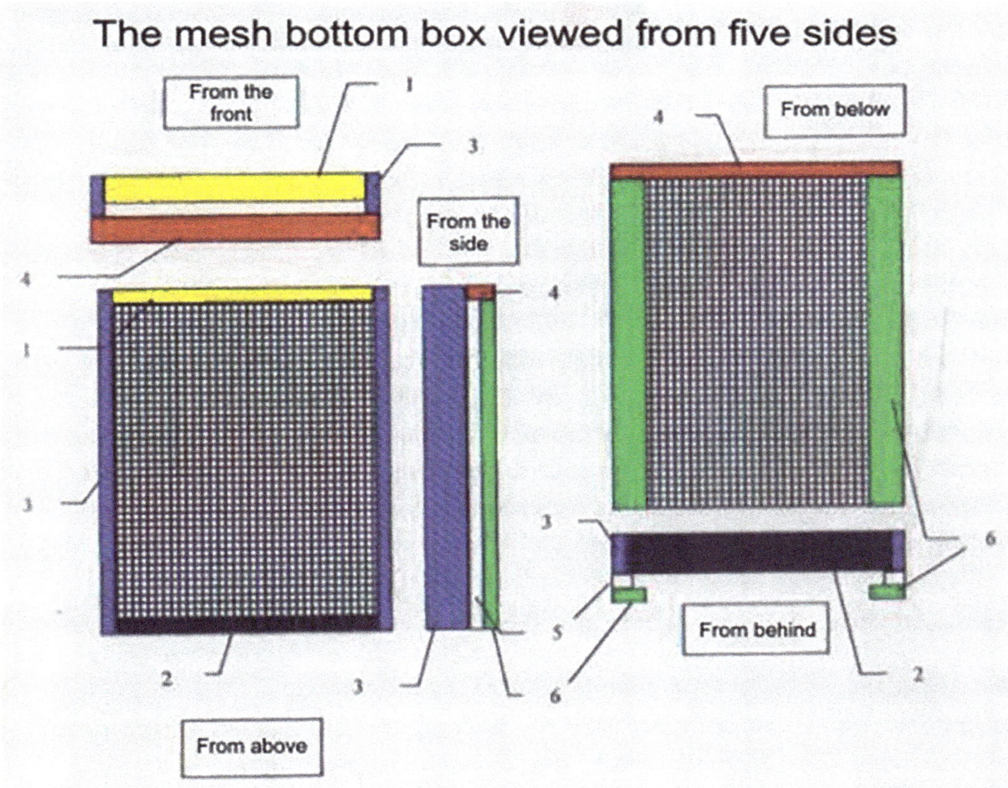

After a drawing by Dr. G. Liebig

No	Designation	Number	Length (mm)	Width (mm)	Thickness (mm)
1	tcp crossbar, front	1	380	40	20
2	tcp crossbar, back	1	380	40	20
3	tcp length bar	2	520	60	20
4	bottom crossbar, front	1	420	40	20
5	bottom longitudinal bar	2	500	30	20
6	bottom sliding slat	2	500	50	20

First, the upper "window" is assembled from slats 1, 2 and 3. These slats have the same thickness and, with the exception of slat no. 1, they are the same height. The lower height of slat no. 1 is the height of the flight gap (the flight hole). In summary: slats no. 2 and no. 3 are 60 mm high, slat no. 1 is only 40 mm. So the flight gap has a height of 20 mm. I screw and glue these slats together, but it is of course also possible to connect them with swallow joints. Something for the seasoned professional, I suspect. It is not necessary. Those who like to work carefully can pre-drill the screw holes. In any case, care must be taken to ensure that the slats are connected to each other at right angles.

The assembly of the lower "window" begins with the stapling or at least fixing of the gauze to the front crossbar no. 4. In the finished gauze bottom box, this bar is the bottom of the flight gap. Bar no. 4 is placed above bar no. 1 on the upper "window" and the gauze is then fixed to the long bars no. 3. The previously glued bars no. 5 and no. 6 are then placed on bar no. 3 and fixed. It must be ensured that the nails or screws at the flight gap do not hinder or make impossible the subsequent front attachment.

For the waste diagnosis (counting the fallen varroa mites), a somewhat laboriously constructed inspection board can be used. Such an inspection board consists of a plastic plate with four glued and/or screwed-on edge bars. However, it is more convenient to buy a ready-made, complete plastic inspection board.

The edge also serves to prevent the fallen mites from being blown away by the wind. In case of emergency, a hardboard or plywood board can also be used with a suitable slat on the back, so that the inspection board is still somewhat closed. Counting the fallen varroa mites works best and easiest if the inspection board is white or painted white.

Since I once made about twenty mesh bottom boxes at the same time, without having taken photos of the 'construction process', I will limit myself to the description above with the photos of a mesh bottom box that has already been made. Nevertheless, the photos reveal once again how the mesh bottom box can be made. The mesh used in that mesh bottom box is aluminium mosquito mesh (mosquito netting), which breaks easily and is not mouse-proof. If I were to build new mesh bottom boxes, I would use (stronger) mesh with a slightly larger mesh size, so that I know for sure that all varroa mites are also on the inspection board. Practice makes perfect...

mesh bottom box seen from above

mesh bottom box seen from below

mesh bottom box seen from the front

mesh bottom box seen from the side

mesh bottom box seen from behind

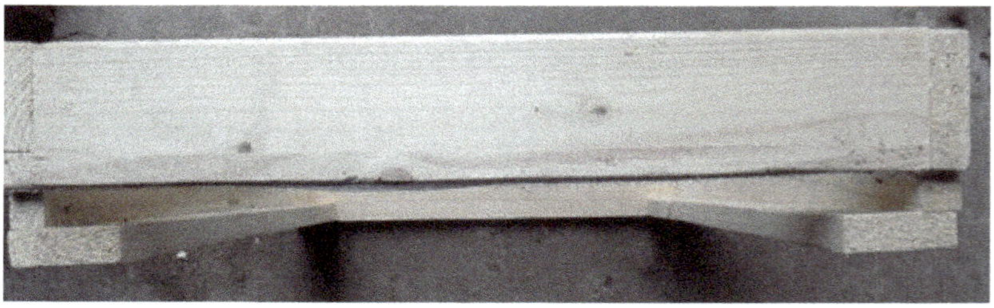

mesh bottom box from below - front

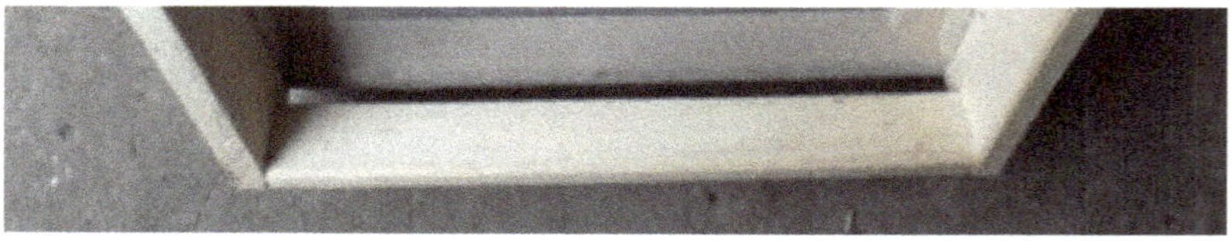

mesh bottom box from bottom - back

Lid (inner lid)

A hive has two lids: an inner lid and an outer lid = the roof.

The (inner) lid consists of a plank (plywood, hardboard) on which four slats are attached to the ends.

The (outer) lid is also the roof. It can be made of wood, but it is more convenient and durable to buy a metal (tin) roof.

The wooden parts of the lid and their dimensions for a Liebig/Zander box.

Designation	Number	Size (mm)	Thickness (mm)
Width slat	2	380 x approx. 20-40	approx. 20
Length slat	2	520 x approx. 20-40	approx. 20
Plate	1	520 x approx. 420 (416, depending on the thickness of the box [hive] walls)	approx. 2-10

Before the lid is placed on the top box or honey box, a sheet of plastic foil is first placed. The foil ensures that the bees cannot glue the lid with propolis and it also makes it possible to look into the beehive without disturbing the bees. Incidentally, the foil does not affect the water content of the honey when it is on top of the honey chamber.

Time and again there are beginning beekeepers who use a perspex plate instead of a plastic foil. Only when that plate is completely glued and difficult to remove without damage do they learn that it is better to use a plastic foil.

Roof

metal roof - top

metal roof - bottom

Wooden roof:

In a 'romantic' whim I had hives with unpainted wooden roofs on an apiary for a few years. After a while those roofs started to get a bit rickety and had to be restored. But below are just a few pictures of such a simple wooden roof. It can be useful in 'emergencies'. The roof shown can also be improved on many points or made more weatherproof.

wooden roof; front or back

top detail >>>

wooden roof; top

wooden roof; bottom

After the front and back of the roof have been drawn (in the angles of choice, but at least 10% difference in size between both ends) on a plank and have been sawn out, the two planks that have also been sawn to size are glued and screwed. Against the worst weather influences, the groove at the top is covered with a screwed-on slat.

Building a three-frames hive, a manual

Based on a solid pine panel of 18 mm thick x 250 mm wide x 2000 mm long.

The panel is sawn (for fitting Zander size frames) into and the remaining panel can already be cut into a piece of 52 x 23 cm (which is needed for building a brood box or honey box). All other leftover wood (take with you and) keep.

- 2 pieces of 52 cm long
- 2 pieces of 12 cm long and 23 cm wide
- 2 pieces of 12 cm long and at least 4 cm wide
- 2 pieces of 12 cm long and 2.5 -3 cm wide.*
- 2 slats or pieces of wood of 15 x 4 cm

* I usually make the last four (2 + 2) pieces of wood myself from leftover wood or other pieces of wood.

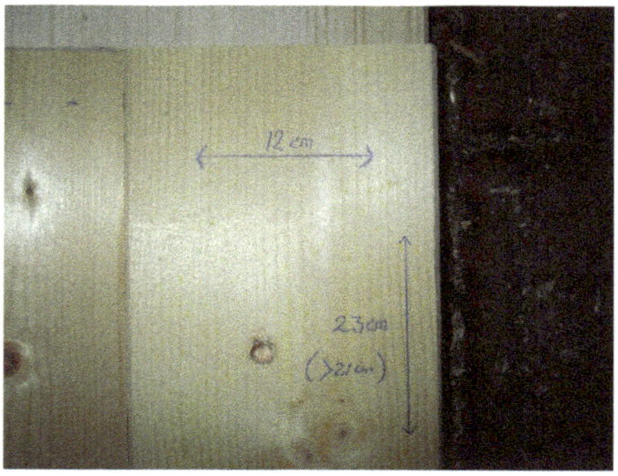

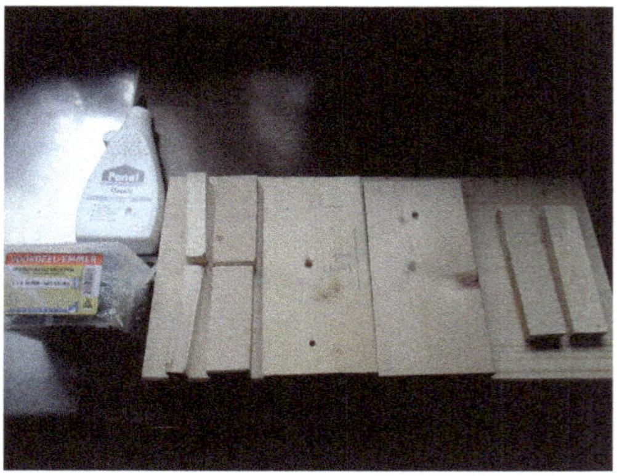

- piece of insect screen (mosquito mesh) of 52x16 cm made of aluminium or steel
- and also needed:
- screws (3.5 x 30 mm, flathead -best stainless-),
- nails (1.6 x 30 mm -best stainless-),
- wood glue,
- a screwdriver,
- a hammer,
- a stapler,
- (a pair of scissors or cutting pliers) and
- a wood drill

We take the two pieces of wood of 2 cm length and 23 cm width and also the two pieces of wood of 2 cm length and -at least- 4 cm width that all seemed to be left over as leftover pieces when sawing.

A front or back wall consists of a larger piece of wood and a smaller piece of wood that are glued and nailed together.

(See also already the photos below.)

On a piece of wood 12 cm wide and 23 cm long a strip of 12 cm and -at least- 4 cm is glued and nailed. The total length of these two pieces of wood together should be 25 cm.

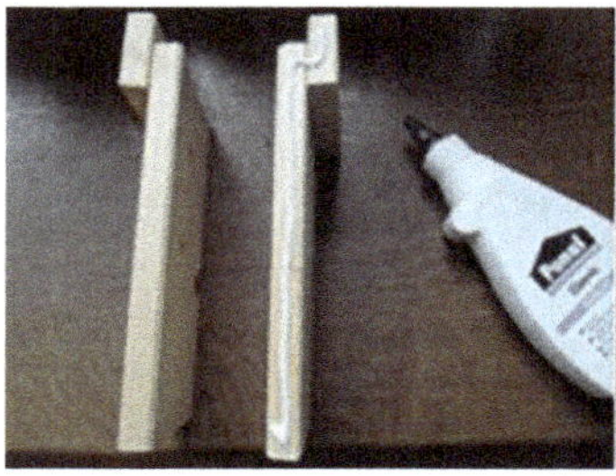
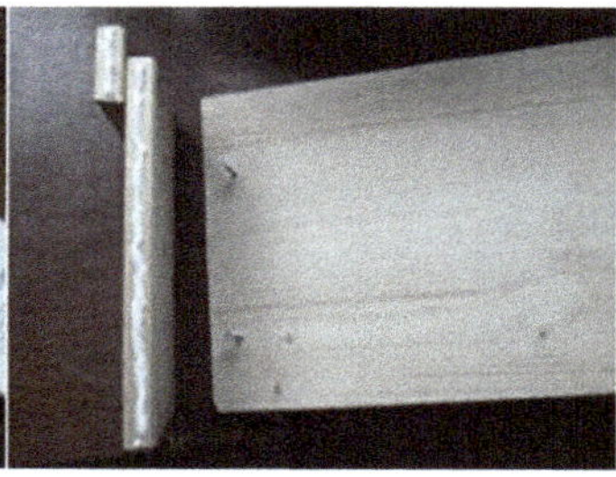

Now follows the assembly of the front or back wall on a side wall.

First I turn all the screws a little into the side wall. Then I put some glue on the side of the front or back wall in progress. Then I screw the side wall on.

The other sides are connected to each other in the same way.

If this has not already been done, the small, narrow pieces of wood are glued to the front and back walls (which are still identical) and nailed to the side walls.

We can now clearly see what the bottom of the hive is (you can't hang your frames in there). We're going to attach the mosquito netting (mosquito mesh) to the bottom. If we use aluminum mosquito netting, we cut it to the right size with scissors. You can also first staple the mosquito netting and then trim it if necessary.

Finally, so far, the two pieces of wood on which the cabinet will stand are nailed to ensure sufficient ventilation through the mosquito mesh.

Now it is decided which will be the front and which the back of the box. With a wood drill with a thickness of 10 mm, two holes are drilled directly next to each other on the bottom side, slightly to the left or right of the middle of the front wall. It is best to on choose a place on the side where the brood frame will later hang in this little split hive which the honey bees will make the brood cells.

The two holes can now be filed down a little to form an oval (through which a maximum of two bees can pass).

The three-frames box is now ready. There is still a lid missing. For that I use a plywood plate that fits on the top. When using the box with bees I simply stick that lid with painter's canvas.

Four compartment bottom box:

www.ingramcontent.com/pod-product-compliance
Ingram Content Group UK Ltd.
Pitfield, Milton Keynes, MK11 3LW, UK
UKHW061957041026
15591UKWH00016B/1046

9 781913 811198